Lecture Notes in Computer Science   14336

Founding Editors

Gerhard Goos
Juris Hartmanis

Editorial Board Members

The series Lecture Notes in Computer Science (LNCS), including its subseries Lecture Notes in Artificial Intelligence (LNAI) and Lecture Notes in Bioinformatics (LNBI), has established itself as a medium for the publication of new developments in computer science and information technology research, teaching, and education.

LNCS enjoys close cooperation with the computer science R & D community, the series counts many renowned academics among its volume editors and paper authors, and collaborates with prestigious societies. Its mission is to serve this international community by providing an invaluable service, mainly focused on the publication of conference and workshop proceedings and postproceedings. LNCS commenced publication in 1973.

Ruben Rios · Joachim Posegga
Editors

Security
and Trust Management

19th International Workshop, STM 2023
The Hague, The Netherlands, September 28, 2023
Proceedings

Springer

Editors
Ruben Rios ⓘD
University of Malaga
Málaga, Spain

Joachim Posegga ⓘD
University of Passau
Passau, Germany

ISSN 0302-9743   ISSN 1611-3349 (electronic)
Lecture Notes in Computer Science
ISBN 978-3-031-47197-1   ISBN 978-3-031-47198-8 (eBook)
https://doi.org/10.1007/978-3-031-47198-8

This Springer imprint is published by the registered company Springer Nature Switzerland AG
The registered company address is: Gewerbestrasse 11, 6330 Cham, Switzerland

Paper in this product is recyclable.

Preface

The International Workshop on Security and Trust Management (STM) has a long history, with great pleasure we present here the proceedings of the 19th International Workshop on Security and Trust Management held in September 2023 in The Hague, Netherlands. STM 2023 was conducted in conjunction with the 28th European Symposium on Research in Computer Security (ESORICS 2023), marking another milestone in the workshop's illustrious history.

As one of the pillars of ERCIM's Security and Trust Management Working Group, STM remains committed to fostering and catalyzing innovative research across a spectrum of subjects, including cryptographic protocols, identity management, security and trust metrics, privacy and anonymity, social implications for security and trust, and the interplay with novel technologies and scenarios.

In response to the call for papers, we received a total of 15 submissions. Each submission was reviewed by at least 3 reviewers, who evaluated the submitted papers based on their significance, novelty, and technical quality. After the reviewing process and a discussion round, only 5 papers were selected as full papers, representing an acceptance rate of 33%. Moreover, given the remarkable quality of the submissions, we enriched the event's program with four succinct yet impactful short papers.

Security and Trust management is a comparably broad area, so contributions are sometimes not easily clustered into sessions dedicated to certain topics; this year, we had papers dealing with identities, application-oriented topics, and privacy-related contributions; these are the three sessions of our 2023 workshop in The Hague.

The workshop program was completed with the traditional talk of the ERCIM WG STM Best Ph.D. Thesis Award. Among the excellent candidates, on this ocassion the award went to Juan E. Rubio for the outstanding contributions made in his doctoral thesis entitled "Analysis and Design of Security Mechanisms in the Context of Advanced Persistent Threats Against Critical Infrastructures". The award was collected on his behalf by Cristina Alcaraz, co-advisor of this thesis together with Javier Lopez.

The success of STM 2023 depends on a long list of individuals who also devoted their time and energy, and provided active support to the organization of the workshop. We would like to thank all the members of the Program Committee and the external reviewers for their collaboration in reviewing manuscripts and selecting the ones with substantial contribution to the thematic area of the workshop. We also gratefully acknowledge all people involved in the successful organization process: the chairperson of the ERCIM STM Working Group, Pierangela Samarati, for her constant support; the ESORICS General Chairs, Kaitai Liang and Georgios Smaragdakis; and the ESORICS Workshops Chairs, Jérémie Decouchant and Stjepan Picek. Last but not least, special thanks to Talaya Farasat for her efforts as Publicity Chair and her continuous support during the preparation of the proceedings.

Last but not least we are also very grateful to the authors for submitting their excellent research results and to all attendees who honored us with their presence and contributed to valuable discussions. We hope that the workshop proceedings will be helpful and inspiring for future research in the area of Security and Trust Management.

September 2023   Joachim Posegga
                                                           Ruben Rios

Organization

Program Committee Chairs

Joachim Posegga   Universität Passau, Germany
Ruben Rios   University of Málaga, Spain

Publicity Chair

Talaya Farasat   Universität Passau, Germany

Program Committee

Cristina Alcaraz	University of Málaga, Spain
Joonsang Baek	University of Wollongong, Australia
Mauro Conti	University of Padua, Italy
Said Daoudagh	ISTI-CNR, Pisa, Italy
Sabrina De Capitani di Vimercati	Università degli Studi di Milano, Italy
Carmen Fernandez-Gago	University of Málaga, Spain
Olga Gadyatskaya	Leiden University, The Netherlands
Dieter Gollmann	Hamburg University of Technology, Germany
Marko Hälbl	University of Maribor, Slovenia
Omar Ibrahim	Hamad Bin Khalifa University, Qatar
Chenglu Jin	Centrum Wiskunde Informatica, The Netherlands
Panayiotis Kotzanikolaou	University of Piraeus, Greece
Hiroaki Kikuchi	Meiji University, Japan
Kwok-Yan Lam	Nanyang Technological University, Singapore
Wenjuan Lia	The Hong Kong Polytechnic University, China
Giovanni Livraga	Università degli Studi di Milano, Italy
Fabio Martinelli	IIT-CNR, Italy
Sjouke Mauw	University of Luxembourg, Luxembourg
Weizhi Meng	Technical University of Denmark, Denmark
Chuadhry Mujeeb Ahmed	Newcastle University, UK
Martín Ochoa	ZHAW and ETH Zurich, Switzerland
Davy Preuveneers	KU Leuven, Belgium
Pierangela Samarati	Università degli Studi di Milano, Italy
Qingni Shen	Peking University, China

Yangguang Tian	University of Surrey, UK
Hiroshi Tsunoda	Tohoku Institute of Technology, Japan
Chia-Mu Yu	National Chiao Tung University, Taiwan

Additional Reviewers

Sergiu Bursuc
Ziyao Liu
Gabriele Orazi
Pier Paolo Tricomi

Contents

Identities

Impact of Consensus Protocols on the Efficiency of Registration and Authentication Process in Self-sovereign Identity

Lydia Ouaili[1,2]([envelope]) [ID]

[1] Conservatoire national des arts et métiers, Paris, France
lydia.ouaili@lecnam.net
[2] Trasna Solutions (Safe-IoT), Marseille, France

Abstract. Self-Sovereign Identity (SSI), allows each entity to control its identities with minimal data disclosure to ensure privacy, through its concepts of Decentralized Identifiers (DIDs) and Verifiable Credentials (VCs), and Zero-Knowledge Proof (ZKP) protocols. The trust system is based on distributed ledger technologies such as blockchain, that provide a tamper proof and correct record of data. Registration and authentication in the SSI depend on the read-write operations that occur on the distributed ledger. In this work, we analyze the impact of distributed protocols on the read/write operations through various efficiency metrics, and therefore, take a step towards deciding which protocols can be suitable for the decentralized identity system. Moreover, by analyzing the properties of consensus algorithms, we propose an efficient protocol for the read operation adapted to the SSI requirements.

Keywords: Self-Sovereign Identity · Decentralized Identifiers · Authentication · Blockchain · Consensus protocols

1 Introduction

The rules and processes involved in Internet Identity Management Systems have always been evolving, due to the challenges of technological innovation [1]. Current digital interactions use identifiers such as user IDs, email addresses and urls, which can be associated with biometric credentials and access tokens. The fact that several services are digitized, interactions require several identifiers with different passwords and privacy policies for each web service. This type of service access fits into the centralized model, which becomes a challenge to manage, because it requires the memorization of logins and passwords and has privacy issues since the service provider stores and controls its users's data.

To overcome the drawbacks of the centralized model, the federated model has emerged and is currently one of the most widely used. This model involves an identity provider, a three-tiers entity that intervenes in the middle of the interaction between the user and the service provider, enabling access to services with a single identity, managed by the identity provider. Certainly, this solves the number of passwords to remember and facilitates interactions between individuals and organizations in the web, but still allows several companies to hold the user's data, and the identity provider tracks all the user activity on the web each time the user makes an authentication.

When Bitcoin was created [2], several decentralized applications have emerged. Their success led to a significant research improvement in distributed and financial system communities. In the financial world, it is possible to transfer a value anonymously without any central authority. A bit later, in 2015, blockchain technology was presented as the solution to privacy and security issues in current identity management models (centralized and federated models) [3]. The need for an entity to have a total control over its identifiers and authenticates on the web without a central authority has emerged. This is one of the issues addressed by the Self-Sovereign Identity (SSI) with its (DIDs) and (VCs), enabling entities to divulge only the necessary information by ZKP protocols with the possibility to revoke them at any time. The trusted system mimics what happens in the real world, it considers three entities: the holder, the verifier and the issuer. These entities all trust and rely on the distributed ledger, to create DIDs, seen as a registration process and issue VCs (done by the issuer) to prove assertion about the holder but also for authentication, to access the verifier services.

Although the fundamental concepts of SSI are independent of Blockchain technology, they are combined together in the trust model. Blockchain technology enables distributed protocols to manage and store unique identities and avoids the need for a trusted third party (TTP) [4]. However, centralization is not the only aspect, since conventional traditional identity management systems (e.g. OAuth, OpenID) based on accounts and digital certificates (X.509 standard) pose a privacy problem. Certificates contain sensitive information, which compromises the privacy of certificate holders, and it might not be appropriate to store them in an immutable distributed register that can be read by several entities. To solve these problems, the DID has been proposed to identify anonymously the holders, but also to be associated to the DIDdocument, which will replace the certificates and contain only encrypted information to authenticate the holders.

Although the conceptual parts of the SSI and their link with Blockchain have been addressed by many studies, there is a lack of in-depth work on SSI and decentralization based on distributed ledger technologies. In the financial model, a similar issue received a significant attention. Indeed, Blockchains depend mainly on consensus algorithms [4], which are required to ensure the management of the ledger. The consensus protocols depend on the use case, which will influence the choice of the type of the ledger (permissioned or permissionless), the security and the integrity of the ledger and other several factors as scalability and throughput and latency.

In this work, we analyse in depth, the properties of consensus protocols, their efficiency metrics in the context of SSI. We highlights that some properties are fundamentals to the verifier and issuer and other properties are optional.

Starting from that, we propose an efficient protocol which first prioritizes the issuer and verifier needs to authenticate the holder.

This paper is organized as follow, Sect. 2 focuses on research's works on SSI from the literature. In Sect. 3 the main concept of SSI are defined. In Sect. 4, we explain the DID creation process, we give the problem definition and our contribution. The last Sect. 5, is dedicated to the proposed protocol.

2 Related Work

There are many research works on the SSI that generally involve an exhaustive study of the concepts, the challenges and benefits of SSI and use-cases.

The first ideas behind SSI can be found in the Sovrin Foundation [5] paper and the vision of Christopher Allen[1]. They recall the chronological evolution of identity management on the web, and consider SSI as the latest model that covers the security and privacy needs of digital interactions.

When it comes to the approach taken to analyze and design SSI components, there is a gap between industry, independent foundations and academic research. The first two actors focus on open source code and software architectures for implementing SSI as a service. Although the open source code of the proposed design, they lacked an in-depth analysis and a formal framework on the different solutions to explain particular implementation choices such as the choice of the distributed ledger, this led to many research studies that present a formal framework and a complete analysis of the major concepts (DIDs and VCs) [6–10], as well as its compatibility with the General Data Protection Regulation (GDPR) [12,13]. Other works focus on DIDs and VCs creation and trust model for authentication [19], discussing the difficulties and limitations of these concepts [14,15]. Analysis and comparisons of Sovrin and uPort implementations based on SSI requirements (e.g. longevity, security privacy, interface, scalability) have been proposed in [16]. In [17], authors proposed architectures and design models for software development.

Blockchain has been seen as a good candidate for overcoming drawbacks related to classic identity management models [20–22]. They are generally non-transparent with regard to their privacy policies and implementations. In addition, certification is costly and controlled by certification authorities (CAs).

In terms of use cases, an identity management system for eGovernment has been proposed in [23], where it is possible to move and connect the government's centralized and qualified identities to the SSI model by converting the data format to match the SSI format and moving to the decentralized management using Hyperledger Indy, via an agent acting as an interface. In [10], the authors replace the classical model of resource access control, where storage of sensitive data such as user attributes is required to enable data access by associating VCs and the distributed registry with access policies. In [24], the authors consider the revocation process, which is another fundamental aspect of identity, proposing a model that enables offline revocation in the SSI, that matches real-life scenarios.

---

[1] www.coindesk.com/markets/2016/04/27/the-path-to-self-sovereign-identity/.

The application of the SSI in the IoT context is also considered in several works, technologies and challenges associated with the application of SSI in the IoT context [28]. The authors provide a detailed comparison of the most promising IoT-oriented frameworks (Hyperledger Indy, uPort, BlockStack, VeresOne, Jolocom). The challenges highlighted by this article illustrate the key issues for the adoption and use of SSI in IoT-enabled domains. In article [29], the authors provide an analysis and a comparison of all the identity models (PGP, X.509, SSI) in terms of identifier uniqueness, centralized management, service endpoints, and point out that only SSI enables semantic schemas. They also discuss the benefits of SSI in the IoT context, such as privacy and decentralization. Other challenges are addressed, such as the storage capacities of IoT devices, limited processing and energy resources for cryptographic algorithms widely used in SSI. These constraints arising from the nature of the IoT environment have been considered in [37], and the authors propose another type of distributed ledger recognized as scalable, called the Tangle as an alternative to Blockchain and suitable for the IoT. Unfortunately, the security properties of the Tangle are limited.

Several SSI-based use-cases in the IoT context have been considered, for identifying devices [32] by creating and storing DIDs. In [35], the authors consider a use case of the Internet of Vehicles. Their model identifies vehicles with DIDs and records the emission rates produced by the vehicles in the Blockchain to provide verifiable and non-corrupted data. Typically, proof-of-concepts for identity management systems based on SSI are implemented with permissioned Blockchains, such as Hyperledger Indy [2], dedicated exclusively to decentralized identities.

Regarding the privacy, benefits and limitations of using DIDs are addressed in [27] and more recently in [26], the authors proposed a privacy-preserving authentication system using SSI for electric vehicle charging.

In this paper, we focus on the decentralization layer and the distributed system part, to analyze various efficiency metrics that impact the DIDs creation, the authentication process and privacy requirements.

3   Preliminaries

In this section, we recall the main concepts of SSI, as well as the general steps of authentication process, which depends on the Blockchain.

Decentralized Identifiers. At a conceptual level, a DID is a type of persistent identifier. It differs in several levels from traditional identifiers such as domain name systems. It identifies any type of entities (human, organization, website, IoT devices, etc.). It is created in a decentralized way and is cryptographically verifiable so it can be controlled by the entity using it if it owns the public/private key associated to the DID. It is possible to create several DIDs to ensure that entities are not tracked. At a functional level, the DID is created by the network that manages the distributed ledger with a specific method to ensure the uniqueness of the DID, its persistence on the Blockchain and its association

---

[2] hyperledger.github.ioindy-did-method#about.

with public/private key of its owner. The Blockchain will be used to resolve the DID by displaying a *DIDdocument* containing metadata such as public keys, authorizations, timestamps, expiration date, etc.

Verifiable Credentials. On the web, there are a variety of services, some of which allow you access with an anonymous login, while others require additional information about your identity (age, surname, first name, citizenship, etc.). In traditional identity management models, this information is provided via a form that is filled in or scanned identity card and stored by the service or identity provider. In SSI, there is a real paradigm shift. Since the DID does not depend on any central authority, VCs are associated with certificates that prove assertions about the entity controlling the DID. For example, if she/he is an adult, a citizen of a country, etc. VCs are signed by a trusted authority and are used in an authentication mechanism. They assert statements without disclosing sensitive data using ZKP protocol. The DID holder shows the VCs to the verifier to access the services and, through the Blockchain, the verifier confirms that the digital signatures are valid.

Zero-Knowledge Proof. For privacy enhancement of SSI, VCs are linked to ZKP [36]. With ZKP, it is possible to prove statements such as "I have a signature", without saying anything additional (i.e. without revealing what is the associated value of the signature). For a holder to use VCs with ZKP, an issuer must sign the claim about the holder, a "proof", so that the holder can present the information to the verifier in a way that enhances privacy. One common practice is to prove the knowledge of the signature without revealing the signature itself.

General Interaction in SSI. The creation of the DID is done via the Blockchain, and the authentication process requires the holder's DID to prove its digital identity (control the DID) and the issuer's involvement if the holder

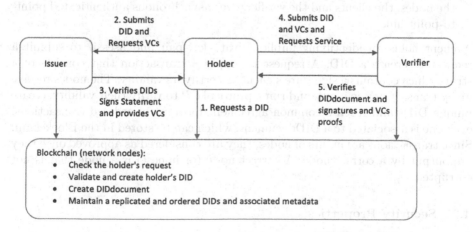

Fig. 1. A high-level workflow of registration and authentication in SSI

has to prove statements. The high-level workflow of registration and authentication is shown in Fig. 1. The holder must first have a DID by initiating a request to network nodes that manage the Blockchain, then it associates the DID with VCs, to present them to the verifier.

4   DID Creation Model and Problem Definition

We first define our DID creation model, in an abstract way, with its associated problem. We focus only on the mechanism to order and confirm the correctness of the set of DIDs created and their associated DIDdocument set that constitutes a block of the Blockchain. This step is fundamental to the registration and authentication process, as the issuer and verifier check that the DID has been created when it is displayed in the Blockchain. To fits with web services glossary and SSI context, in this paper we consider holder and clients synonyms.

4.1   Security Model

We now describe the adversarial model of consensus protocol and trust assumptions related to back-and-forth interaction between the nodes, the verifier and the clients.

- *Adversarial model*: We consider a set $\mathcal{P} = \{1, .., n\}$ of n designated nodes (i,e., physical machines), with distinct, fixed and well-known identities of a private key sk, with a corresponding public key pk. We allow a static adversary in the sense that the adversary can corrupt f nodes (i.e., processes that deviate from protocol specification [38] to compromise the consensus properties) from $\mathcal{P}$, where $f < n/3$ before the execution of consensus protocols (this is a standard assumption in the majority of permissioned Blockchains that rely on a classical consensus protocol [57]).
- *Communication model:* We assume asynchronous communications between the nodes, the clients and the verifier, over asynchronous authenticated point-to-point link.

A client holds an identifying public/private key-pair and uses it to submit a request to create a DID. A request is simply a transaction that consists of a string which contains a signature's client to certify its validity. The nodes receive many transactions as input, and run a protocol $\mathcal{CP}$ to verify their validity, create unique DIDs and reach a common agreement for a set of ordered transactions, each one is associated to a DIDdocument, which can be stored in the Blockchain. Since transactions are inputs of nodes, they are considered as approved once they are output by a correct node. A correct node (or honest) is a node that is not corrupted.

4.2   Security Properties

The consensus protocol $\mathcal{CP}$ outputs a set of approved DIDs (and other metadata related to DIDdocument as public key) such that the following properties hold:

- *Agreement*: If a correct node approves a transaction with $DID = d$ then no correct node approves a $DID = d'$, such that $d' \neq d$ with the same public/private key-pair.
- *Total Order*: If one correct node has output the sequence of transactions $\langle DID0, DID1, ..., DIDj \rangle$ and another node has output $\langle DID'0, DID'1..., DID'j' \rangle$ then $DIDi = DID'i$ for $i \leq \min(j, j')$.
- *Censorship Resilience*: If a transaction associated to a fixed public/private key-pair is input to $N - f$ correct nodes, then it is eventually output by every correct node.

The first property ensures the notion of consistency, in the SSI context, this translates into agreement between network nodes on the uniqueness of the identifier associated with the holder's public/private key. The order in which the transactions are displayed allows the same replicas to be shown. Unlike in the financial world, the order of transactions in many uses cases in SSI is not important for authentication process, as we will explain later. The last property ensures the protocols' liveness, i.e., it prevents the omission of a correct DID request from being approved. Traditionally, a consensus protocol providing such properties for general transactions is called atomic broadcast [55], with modern consensus parlance, we would call it a Blockchain.

4.3 Authentication Process and Read Operation on Blockchain

Note that each existing application services as social-media platform, or institutions platform as online banks, or email-account services, have their own rules and policies for access control, some may require only an identifier as a DID from a specific Blockchain, other may require additional proofs as VCs from different issuers.

According to the general interaction in SSI (see Fig. 1), registration and authentication are carried out incrementally as follows:

- Registration process: *Clients* request a DID to nodes in $\mathcal{P}$.
- Each correct node from $\mathcal{P}$ approves a DID request following some protocol $\mathcal{CP}$ and generates blocks of committed transactions.
- *Clients*: Submit the DID and request VCs from the *Issuer*.
- *Issuer*: extracts DIDdocument from Blockchain and issue VCs
- *Clients*: submit VCs and DID to the *application verifier*.
- Authentication process: *application verifier* extracts and verifies DIDdocument from Blockchain. It ensures that the client controls the public key associated to DID and verifies VCs proofs.

From previous steps, we observe that the DID verification is done twice by the issuer to issue VCs and the application verifier to authenticate clients. DID verification and DIDocument access are done through a read operation in the Blockchain by extracting the associated DIDdocument from an approved block.

In the following, we explain how the read operation is executed on the blockchain. Since issuers, clients and application verifiers read the Blockchain to gather the DID and extract the DIDdocument information, we call them Readers.

A Reader queries nodes from $\mathcal{P}$ through an interface system to obtain a result of a read operation. Recall that we assume the existence of f corrupted nodes (see Sect. 4.1, a standard assumption in the majority of permissioned Blockchain). So querying just one machine is not sufficient to obtain the correct blocks of DID, since Reader may observe conflicting data from f. So to mask the answers of corrupted machines, the read operation is considered valid if at least $f + 1$ gives a matching answer of blocks to the interface system, to be sure that any approved block of DIDs was returned by at least one correct node. So if at least $f + 1$ nodes provide the same block after executing the whole consensus protocol, the block is considered as approved. Consequently, registration and authentication process rely on read operation which requires at least $f + 1$ nodes confirming the same DIDdocument.

4.4 Efficiency Metrics

In what follows, we consider various metrics that impact the performance of the protocol $\mathcal{CP}$, and therefore SSI at the level of read operation after, DID creation step, but also at the other steps (see Fig. 1), since the issuer and verifier consult the Blockchain. We will also explain the benefits of these metrics in two use cases. We consider four metrics of interest:

- Scalability membership: since the nodes will replicate a consistent database, increasing the number of nodes n is important to secure the replicated data as Blockchain and to improve its availability.
- Horizontal scalability: DID requests rate increases as the number n of nodes increases.
- Latency: latency is defined as the time interval between the time the first node receives a client request for creating a DID and when the (n-f)-th node finishes the protocol $\mathcal{CP}$.
- Throughput: Throughput is defined as the number of DIDs approved per unit of time.

Realizing these metrics simultaneously is extremely challenging, scaling the number of participants for Blockchain-based systems is a fundamental problem, since it increases the number of messages exchanged and therefore delays the agreement (Sect. 4.2). Horizontal scalability is desirable when deploying SSI for ubiquitous use. Indeed, as the number of digitized services and users continues to grow, DID creation must be efficient, to provide a reasonable registration latency. If, for example, we use the bitcoin protocol [2] in an open membership (anyone can participate to the protocol, join and leave the network), we have to wait 10 min to create and confirm a DID. However, this is not the case of byzantine fault-tolerant systems (BFT) [39] since one can achieve 10,000 transactions (TX) per second (s) if $n < 16$ [40]. But, if we increase n to 64, the throughput decreases drastically and one can achieve 5000 TX/s [40]. Therefore, the membership scalability is not possible with BFT, which makes the BFT unsuitable for the open Internet but possible in permissioned Blockchain [41] like Hyper-ledger Indy or Hyperledger Fabric. To overcome the scalability issues of the previous work, sharding techniques (in open-membership) [42–44] are promising solutions to achieve horizontal scalability, where there is a multiple committee and multiple sets of transactions allowing each committee to manage a fixed set of transactions. In [44], one can achieve 7800 TX/s with $n = 4800$ and 8.7 s of latency. The major drawbacks of sharding protocols is the fact that they are optimal only if transactions belong to the same committee [45].

In our work, we consider two general use cases involving SSI, in terms of using DID for authentication:

- The first use case, is to use the same DID to access applications services. In this situation, if we choose a Blockchain with a good scalability but with high throughput and latency, we would tolerate the latency of creating a DID only once. Then, each time a DID is used, the authentication consists of verifying that the client controls the DID through cryptography and the time it takes to consult the DIDdocument in the Blockchain and proofs of VCs. On the other hand, using the same identifier raises privacy concerns. Indeed, correlation attack may occur by identifying some patterns behavior of the clients. In traditional identity management models, according to an American report one can identify uniquely US population from multiple attributes as gender and date of birth [46]. Moreover, scenarios of privacy breaches, such as those that happened to AOL [48] and Netflix [47] are concrete examples that trigger the need to questioning anonymisation techniques to enhance and protect privacy [49]. Therefore changing the DID for each authentication would be desirable, to avoid making similar mistakes in SSI context and this is already considered in IoT-based SSI mainly in Electric Vehicle Charging system, where it is possible to track the trajectory of clients and their location. So for privacy enhancement, authors in [27,28] change the identifier for each authentication, which brings us to the next use case.
- The second use case considers a situation where the DID is changed for each authentication. This poses not only scalability problems, as more and more DIDs have to be created, but also latency and throughput problems. In the first case, the holder waits only once for the DID to be created. In this case, the holder has to wait each time it requests the protocol to create a new DID to identify itself.

We saw that the consensus protocol impacts registration and authentication processes efficiency, especially if the DID changes with each authentication, and we also saw that realizing performance metrics simultaneously is a real challenge in the distributed systems and Blockchain context. In this paper, we propose an efficient protocol for the read operation (see Sect. 4.3) used by the issuer and the verifier in the authentication process.

Our identity system lacks certain capabilities compared to previous Blockchain systems as cited above (e.g., sybil attack resistance, SSI-based smart contracts, or SSI-based Hyperledge Indy). Our goal is not to replace these sys-tems but rather to highlight the possibility to design an efficient read operations by considering the protocol $\mathcal{CP}$ properties, agreement and the total order separately.

An important insight in our work, is to highlight, that in previous use cases, the verifier and the issuer need the DIDdocument to extract metadata and a proof that the client controls the DID by reading the output of the consensus protocols which is simply a block containing a valid DID. Consequently, the verifier doesn't need to know all the DIDs present in the Blockchain or the latest DID created but only those who request access to its services. In the next section, this observation will enable us to propose a protocol that reduces the latency of DID confirmation and its creation by analyzing in depth a consensus protocol of [40] that first ensures agreement then total order. The agreement means consistency and that's what the verifier and issuer need to read from the protocol $\mathcal{CP}$. This protocol will be adapted to SSI to improve the latency of read operation to extract the DIDdocument.

5   Proposed Protocol

At the heart of our protocol, lie two building blocks that are distinct from the context. The first concerns the *identity block*, which involves interaction between clients and readers as the verifier or issuer. The second concerns the *protocol block* that creates, validates and stores DIDs in blockchain, through consensus protocol executed by nodes from $\mathcal{P}$. We assume that the client and Readers (issuer and verifier) trust the Blockchain, in the sense that they trust the protocol and they rely on the DID created by this protocol to respectively provide a VCs by issuer and authenticate clients through VCs and DID by verifiers. Note that we are considering an abstract framework, without specifying the details of the application, which includes the design, schema and additional rules, DID identifier and DID document syntax. In our case, a DID creation request is seen as a transaction that is simply a unique string, which will be processed by nodes performing some basic validation and executing a consensus protocol to generate a block of approved transactions. The DIDdocument includes data elements assembled from the transaction. It may contain only DID and public key, or other additional data that depend on rules and policies of the application. Reading the ledger means extracting the DID transaction data into DIDdocument. To simplify the protocol, we consider the read operation model and we don't focus on the next steps as the issuer issues a VCs. Our aim is to improve the verification process which has an impact on the subsequent steps in the interaction of Fig. 1.

5.1   Overview of the Protocol

To explain our protocol, we consider the following assumptions and properties:

– In the *identity block*, clients possess an identifying public/private key-pair before the beginning of the protocol. Readers (issuer and verifier) require the existence of a DID associated to this public/private key and the proof that the client owns it. We refer to this scenario by Client-Readers verification scenario. The DID is provided by the network nodes that manage the Blockchain by checking the validity of transactions to create a unique persistent DID.

– In the *protocol block*, nodes hold a distinct well-known identities (identifying public/private key-pair). They implement an asynchronous BFT protocol of [40] called HoneyBadgerBFT to create and validate DID. The aim of this consensus is to provide the properties cited in the previous Sect. 4, although there are several consensuses that provide the same properties [51–53] except that each protocol has its own critical factors to achieve, e.g. the authors [51,52] focus on minimizing cryptography or steps to reduce latency, but it still remains dependent on timing assumptions [40] and are vulnerable to the adversarial scheduler [40]. In addition, optimizing cryptography can make also the consensus vulnerable to attacks [53]. HoneyBadgerBFT does not care about timing assumptions and still guarantees a good throughput and robustness.

In the modern protocols used in Blockchain, which are mainly used in the financial world, latency is not their critical factor but rather scalability, yet financial institutions as Visa [50] have expressed interest in adopting a high-speed transactions confirmation system.

In the SSI world, we consider latency and throughput to be critical factors for authenticating entities, especially in ubiquitous web-services or IoT context (see uses cases of [27,28]), and so it is important to choose a consensus protocol adapted to SSI needs. The HoneyBadgerBFT guarantees interesting properties for the client-readers verification scenario, it can reach a throughput of 10000 Tx/s, and scale to hundred nodes on a wide area network. In addition, its latency depends on n and for $n = 104$ the latency is 6 min. Note that Hyperledger Indy, uses 25 nodes and is based on RBFT [51]. Moreover, what interests us about this protocol is, that it is the combination of two phases, the first phase satisfies the agreement property and the second one orders the transactions in a block. This is exactly the first step the Readers need to advance in the process of Fig. 1. The second step is optional for them (total order property).

If the client's request is valid, the uniqueness of the DID can be ensured by using the collision-free hash function of its digital signature. For example, in Hyperldger Indy, after verifying the validity of the DID request, the expression of the DID is created by using the hash function of the public key [54].

5.2   DID Creation and Verification Process

In this section and the next one, we recall the important steps of HoneyBadgerBFT protocol, although this belongs to the field of distributed systems, but decentralization is now used in many applications, and a deep understanding of the consensus protocols would considerably improve the application performance. This is already the case in financial institutions such as Visa [44] where a consensus adapted to their needs has been developed. In this section we will adapt HoneyBadgerBFT protocol to our setting (Client-Readers verification scenario Sect. 4.3).

We now describe the three steps involved in creating DID and read operations, and how to adapt HoneyBadgerBFT protocol to SSI.

Submitting a DID Request. In this phase, the client's request consists of creating a message comprising a digital signature. We model this by:

$$\mathcal{R} = \langle client(c), message(m), signature(s) \rangle.$$

We assume that nodes possess an authenticator to verify signature's requests. In the classic SSI model, the client (holder) first waits to obtain a DID, and then submits a request to the issuer and then to the verifier. In our case, we optimize this process by simultaneously informing Readers (issuer and verifier) that the Blockchain nodes are creating the DID and DIDdocument. In the meantime, Readers just check that client owns the public/private key.

Nodes Run HoneyBadgerBFT Protocol. Upon receiving client's requests $\mathcal{R}$, nodes choose a set of requests from their buffer and verify signatures before running the HoneyBadgerBFT protocol. For pedagogical reasons, we will first explain the protocol at a high level, then detail the relevant steps in the next section. The HoneyBadgerBFT protocol is implemented in a modular way. The first nodes encrypt the set of requests for security reasons (to prevent adversary from learning DIDs requests before the $n - f$ corrects nodes) and then implement two protocols: Byzantine Reliable Broadcast (BRC) and Byzantine Agreement (BA). The goal is to agree on a subset of proposed encrypted set of requests. The BRC phase provides a *consistency* property, i.e., if a correct node receives the DID request then every correct node receives the same DID. It ensures that no two correct nodes approve a different DID for the same public/private key. The second phase BA, consists in voting using consensus protocol if the proposed encrypted set will be in the next block, if it is the case, the third phase consists in decrypting all the selected encrypted sets containing the entire requests transactions chosen by at least $n - f$ who have completed the first and second phases. Finally, nodes order DID through canonical method and add them to the block.

We deduce that if a client executes correctly the request (by following the submitting phase), each of HoneyBadgerBFT phase is guaranteed to terminate. If we follow the classical read operation, for Readers to confirm that the DID has been created, they check whether the block containing DID is approved by at least $f + 1$ nodes. In our case, the readers rely on the third phase that concerns the decryption step of selected sets that will appear in the block, since the set proposed by a correct node and selected by a byzantine agreement will be in the next block. Hence, Readers do not have to wait for the block of all transaction to appear in the Blockchain, but focus on specific DIDs of their clients. This optimizes latency for authentication process since our read operation finishes before the blocks appear.

Approving a DID. Let d be some correct DID requests and let i be some correct nodes that consider d in the set S of the next epoch of the protocol.

If at least $f + 1$ nodes decrypt S and thus d in the third phase then the DID associated to d is considered as approved and will appear in the next block (this is explained in the next section).

Clients and Verifier Notification. We assume the existence of an interface which informs the clients and all Readers allowed to read execution dynamics and results, all the approved DIDs and the Blockchain. Readers consider the DID valid if it has been approved by the *approving DID* step. Note that the previous step is executed for all the proposed DID requests of at least $n - f$ nodes, and all the correct nodes need to wait $f + 1$ messages to decrypt each set. In the last phase, a block appears only if a correct node receives at least $(f + 1)$ messages of all the proposed sets from at least $(n - f)$ nodes and thus $(f + 1) \times (n - f)$ received messages are necessary for the block to appear. On the other hand, if we focus only on a particular DID as d, we need $(f + 1)$ nodes that received $(f + 1)$ messages to decrypt d.

5.3   HoneySSIBFT Protocol

In this section, we focus on building a protocol based on HoneyBadgerBFT layer and SSI context, that we call HoneySSIBFT protocol, we give a sufficient conditions for Client-Readers verification scenario and prove that verification process latency is reduced.

Building Blocks:

- **STEP 1.** The process follows a protocol that operates in epochs. At the end of each epoch, a new set of DIDs is added to the committed log. At epoch r, nodes receive a DID's request from multiple clients as input and store them in their buffer. Each node i chooses a set S_i of DID's requests from its queue. For scalability reasons, the size of S_i is B/n where B is a batch size to ensure that each node proposes a distinct set of DID's requests (see [40] for more details on the parameters). Recall that one has f corrupted nodes. Moreover, assuming an asynchronous network, the delivery schedule of messages is entirely determined by the adversary [40]. It doesn't concern SSI explicitly but the protocol take them into account to create consistent blocks of DID.

  To ensure that the adversary is not aware of which transaction is being validated, authors propose to first encrypt the proposed set of messages with tpke primitive $tpke(pk_i, S_i) = C_i$ where $i \subset [1, ..., n]$ (defined below), and then each correct node i responsible for the set C_i, reliably disseminate C_i using RBC to ensure that all correct nodes receive the same C_i. The second phase consists in agreeing through a BA protocol to decide on a binary decision 0 or 1, which means whether or not the set C_i will be added to the next block of the Blockchain. The combination of RBC and BA allow an agreement on a common subset assembled from all the proposed set C_i of at least $N - f$ nodes. Note that C_i is the encrypted message of a set of DID's request S_i. If decision 1 was decided for C_i, then the next step is to decrypt C_i. Note that the *tpke* primitive allow decryption only if each correct node receives an encrypted share value from at least $f + 1$ nodes. The tpke primitive provides the following results:

- For $i \subset [0, ..., n]$, each node i with pk_i encrypts S_i, and gets $tpke(pk_i, S_i) = C_i$.
- $\sigma_{i,j} = tpke.decshare(sk_j, C_i)$ provides an encrypted share value of nodes j on a set C_i (the set is proposed by node i).
- $tpke.dec(pk, C, \{i, \sigma_i\}) \rightarrow S_i$ is a decryption process that requires a value share σ of at least $f + 1$ nodes. So decryption of C_i requires the collaboration of at least $f + 1$ nodes that share their σ values associated with C_i.

In what follows, we adopt the SSI concepts at certain levels that are relevant to Readers (issuer and verifier). When nodes choose S_i they inform the clients with a digital signature that their DID's requests belong to S_i. Each node performs a cryptographic tpke primitive to encrypt S_i by $tpke(pk_i, S_i) = C_i$ and disseminate C_i reliably then runs a consensus protocol to agree on each C_i. The decryption process is done at the end. It reveals the transactions that will be added in the next block. In tpke primitive, the decryption requires the collaboration of nodes by sharing $\sigma_i = tpke.decshare(sk_i, C_i)$. When at least $f + 1$ nodes share their σ_i, it is possible to get the text S_i. Roughly speaking, nodes agree first on encrypted data and then reveal it by $tpke.dec(pk, C, \{i, \sigma_i\}) \rightarrow S_i$.

After encryption, each node i reliably disseminates C_i using byzantine reliable broadcast RBC_i. The set C_i is an encrypted message of client's requests with their signatures. Node i that proposes C_i in RBC_i plays a role of a sender.

The result of executing RBC_i protocol by i is manifested by the event $Deliver(C_i)$. This ensures that the following properties of RBC_i are satisfied:

- (Agreement) If any two correct nodes deliver C_i and C_i'. Then $C_i = C_i'$.
- (Totality) If any correct nodes deliver C_i then all the correct nodes deliver C_i.
- (Validity) If the sender is correct and inputs C_i then all correct nodes deliver S_i.

There is a rich research work about RBC protocols [56,57]. We refer to [40] for the details of the protocol used in HoneyBadgerBFT. This a powerful abstraction to reliably disseminate a message and inform nodes about the encrypted set C_i that contains DID's request. Note that this layer is executed by each correct node with a specific C_i. A reliable dissemination of a correct node allows us to receive a correct message and be sure that all the correct nodes have received it.

- **STEP 2.** The second step of the protocol is an implementation of Binary Agreement (BA) to agree by $b \in \{0, 1\}$, where $b = 1$ in BA_i means that the set C_i of node i will be in the final set and therefore it will be in the next block of the Blockchain. BA guarantees the following properties [40]:
  - (Totality) If any correct node outputs the bit b, then every correct node outputs b.

- (*Termination*) If all correct nodes receive input, then every correct node outputs a bit.
- (*Validity*) If any correct node outputs b, then at least one correct node received b as input.

To sum up theses first steps, nodes choose a set of distinct transactions S_i, they inform clients that there request is currently being submitted to be selected in the next block, They encrypt S_i, then they disseminate reliably the encrypted set by RBC protocol and then they agree whether the encrypted sets will be in the next block by BA protocol. The next step allow to decrypt them and store them in the Blockchain.

- **STEP 3.** The decryption phase is performed for all C_i values that have completed the previous protocols (RBC and BA). For each C_i, node j needs to receive at least $f + 1$ share values $tpke.decshare(sk, C_i)$ to know S_i. After decoding C_i, each correct node performs a basic verification to validate a transaction, e.g., verifies the DID's requests, client's digital signatures, etc. Then, it creates DID through a specific method that depends on the application (one can use the hash function for the uniqueness of the DID) and DIDdocument. We denote by Y_i the associated set of DIDdocument of S_i requests. A correct nodes that decrypt Y_i unicasts to the interface a signed message to indicate that Y_i have been approved. To display transactions, the interface requires $f + 1$ matching Y_i. So, clients, verifiers and issuers are aware of every approved transaction.

- **STEP 4. Add DIDdoc to a block** After decoding all C_i of at least $N - f$ proposed set and creating all DID's data, the next block contains $block_r :=$ $sorted(\bigcup_{i \in E} Y_i), E \subset [1, ..., n]$ such that $block_r$ is sorted in a canonical order and unicasts $block_r$ to the interface. Then, each node removes S_i from its buffer and considers the next DID's requests of its buffer.

5.4   Sufficient Conditions for Read Operation in HoneySSIBFT Protocol

In this section, we focus on the read operation that issuer and verifiers rely on, to respectively provide VCs and authenticate clients, we refer to Sect. 4.3 which explains that receiving $f + 1$ matching block of transactions at the end of protocol epoch execution, means that transactions are considered to be confirmed and each correct nodes will output the same block. In HoneySSIBFT, verifiers and issuer rely on **STEP 3**, to consider a valid DIDdocument. In the following we prove that it is a sufficient condition to verify that client request is approved by nodes and will appear in the block, we prove also that latency of read operation of HoneySSIBFT is less than the classical read operation of Sect. 4.3:

Proposition 1. *Let c be some client, which issued a DID request $\mathcal{R}_c$. Assume that nodes are going to execute the epoch r and assume that $\mathcal{R}_c \in S_i$, where S_i is a set of requests from a correct node's buffer. So, if at least $f + 1$ nodes decrypt C_i (encrypted set of S_i) at **STEP 3** and unicast Y_i to the interface (we denote this instance by $Notification_{f+1}$). Then:*

– $\mathcal{R}_c$ has been selected to be in the next block and its associated DIDdocument is considered as approved.

Let δ and δ_n be respectively the latency of classical read operation (see Sect. 4.3) and the latency of read operation from $Notification_{f+1}$. Then, $\delta_n \leq \delta$

Proof. Since $f + 1$ nodes decrypt C_i in epoch r, then there exists at least one correct process denoted by N_r that execute three steps of HoneySSIBFT **STEP i**, with i = 1,2,3. So, from **STEP 1**, N_r outputs $Deliver(C_i)$ from RBC_i protocol, then any correct process deliver C_i (see properties of RBC protocol), from **STEP 2** N_r outputs 1 from BA_i protocol, then any correct node output 1 from BA_i. Which means that C_i is selected to be in the final set $block_r := sorted(\bigcup_{i \in E} Y_i), E = [1, ..., n]$. Since C_i is a encrypted set of transactions, DIDdocuments are not created yet. This is done in **STEP 3**, N_r received $f + 1$ decryption shares of C_i. So each correct node decrypts C_i and create DIDdocuments of transactions of S_i. This is proves the first property.

To prove the second property, it's enough to analyze the number of steps needed in the classical read operation and read operation with $Notification_{f+1}$. The classical read operation of $\mathcal{R}_c$ requires that at least one correct node commits a block including DIDdocument associated to $\mathcal{R}_c$. To commit a block a correct nodes, wait for at least $(f+1)$ decryption shares of all S_i proposed by at least $n-f$ nodes i, so it waits for $(f+1) \times (n-f)$ messages in **STEP 3**. In $Notification_{f+1}$, DIDdocument of $\mathcal{R}_c$ is approved, if at least a correct process decrypt C_i, creates Y_i and unicasts Y_i to the interface. For that, it is enough to receive $f+1$ messages of decryption shares of C_i.

This result allows us to show that if we're interested in creating a VCs to a specific client and authenticating it, we don't need to wait for the end of the consensus protocol to read the block from the blockchain, but rather focus on the transaction associated with the client's request and the dynamics of the protocol execution. These results are relevant when we have several issuer and verifiers trusting the same blockhain. Clients register via the blockchain but request different services, and clients have a distinct services choice, and each service has its own VCs and issuers policies.

6   Conclusion

In this paper, we focused on authentication and registration in SSI. We only considered the decentralization layer and the distributed system part, to analyze various efficiency metrics that impact the fundamental step of DIDs creation. We highlighted that the choice of consensus is important in the efficiency of interactions in SSI and that some properties are fundamentals to the verifier and issuer and other properties are optional for them. HoneySSIBFT protocol prioritizes the issuer and verifier needs to authenticate clients. It is based on algorithm HoneyBadgerBFT which achieve a reasonable throughput, scalability, latency and satisfies agreement and total order [40]. Our future work consists in conducting an experimental study to analyze the gap between the latency required for blocks to be approved and the latency required for individual transactions to be approved, as suggested in our protocol.

References

1. Bertino, E., Takahashi, K.: Identity Management: Concepts, Technologies, and Systems. Artech House, Canton (2010)
2. Nakamoto, S.: Bitcoin: a peer-to-peer electronic cash system. Decentralized Bus. Rev. 21260 (2008)
3. SBIR. Applicability of Blockchain Technology to Privacy Respecting Identity Management (2015). https://www.sbir.gov/sbirsearch/detail/867797
4. Zheng, Z., Xie, S., Dai, H.N., Chen, X., Wang, H.: Blockchain challenges and opportunities: a survey. Int. J. Web Grid Serv. **14**(4), 352–375 (2018)
5. Tobin, A.: Sovrin: what goes on the ledger? Evernym/Sovrin Foundation, pp. 1–11 (2018)
6. Ferdous, M.S., Chowdhury, F., Alassafi, M.O.: In search of self-sovereign identity leveraging blockchain technology. IEEE Access **7**, 103059–103079 (2019)
7. Naik, N., Jenkins, P.: Governing principles of self-sovereign identity applied to blockchain enabled privacy preserving identity management systems. In: 2020 IEEE International Symposium on Systems Engineering (ISSE), pp. 1–6 (2020)
8. Naik, N., Jenkins, P.: Sovrin Network for decentralized digital identity: analysing a self-sovereign identity system based on distributed ledger technology. In: 2021 IEEE International Symposium on Systems Engineering (ISSE), pp. 1–7 (2021)
9. Čučko, Š, Turkanović, M.: Decentralized and self-sovereign identity: systematic mapping study. IEEE Access **9**, 139009–139027 (2021)
10. Belchior, R., Putz, B., Pernul, G., Correia, M., Vasconcelos, A., Guerreiro, S.: SSI-BAC: self-sovereign identity based access control. In: 2020 IEEE 19th International Conference on Trust, Security and Privacy in Computing and Communications (TrustCom), pp. 1935–1943. IEEE (2020)
11. Mühle, A., Grüner, A., Gayvoronskaya, T., Meinel, C.: A survey on essential components of a self-sovereign identity. Comput. Sci. Rev. **30**, 80–86 (2018)
12. Naik, N., Jenkins, P.: Your identity is yours: take back control of your identity using GDPR compatible self-sovereign identity. In: 2020 7th International Conference on Behavioural and Social Computing (BESC), pp. 1–6. IEEE (2020)
13. Kondova, G., Erbguth, J.: Self-sovereign identity on public blockchains and the GDPR. In: Proceedings of the 35th Annual ACM Symposium on Applied Computing, pp. 342–345 (2020)
14. Brunner, C., Gallersdörfer, U., Knirsch, F., Engel, D., Matthes, F.: DID and VC: untangling decentralized identifiers and verifiable credentials for the web of trust. In: 2020 the 3rd International Conference on Blockchain Technology and Applications, pp. 61–66 (2020)
15. Stokkink, Q., Ishmaev, G., Epema, D., Pouwelse, J.: A truly self-sovereign identity system. In: 2021 IEEE 46th Conference on Local Computer Networks (LCN), pp. 1–8 (2021)
16. Naik, N., Jenkins, P.: uPort open-source identity management system: an assessment of self-sovereign identity and user-centric data platform built on blockchain. In: 2020 IEEE International Symposium on Systems Engineering (ISSE), pp. 1–7. IEEE (2020)

17. Liu, Y., Lu, Q., Paik, H.Y., Xu, X.: Design patterns for blockchain-based self-sovereign identity. In: Proceedings of the European Conference on Pattern Languages of Programs 2020, pp. 1–14 (2020)
18. Grüner, A., Mühle, A., Meinel, C.: ATIB: design and evaluation of an architecture for brokered self-sovereign identity integration and trust-enhancing attribute aggregation for service provider. IEEE Access **9**, 138553–138570 (2021)
19. Nokhbeh Zaeem, R., et al.: Blockchain-based self-sovereign identity: survey, requirements, use-cases, and comparative study. In: IEEE/WIC/ACM International Conference on Web Intelligence and Intelligent Agent Technology, pp. 128–135 (2021)
20. Kaneriya, J., Patel, H.: A comparative survey on blockchain based self sovereign identity system. In: 2020 3rd International Conference on Intelligent Sustainable Systems (ICISS), pp. 1150–1155. IEEE (2020)
21. Bhattacharya, M.P., Zavarsky, P., Butakov, S.: Enhancing the security and privacy of self-sovereign identities on hyperledger Indy blockchain. In: 2020 International Symposium on Networks, Computers and Communications (ISNCC), pp. 1–7. IEEE (2020)
22. Gilani, K., Bertin, E., Hatin, J., Crespi, N.: A survey on blockchain-based identity management and decentralized privacy for personal data. In: 2020 2nd Conference on Blockchain Research & Applications for Innovative Networks and Services (BRAINS), pp. 97–101. IEEE (2020)
23. Abraham, A., Theuermann, K., Kirchengast, E.: Qualified eID derivation into a distributed ledger based IdM system. In: 2018 17th IEEE International Conference on Trust, Security and Privacy in Computing and Communications/12th IEEE International Conference on Big Data Science and Engineering (TrustCom/BigDataSE), pp. 1406–1412 (2018)
24. Abraham, A., More, S., Rabensteiner, C., Hörandner, F.: Revocable and offline-verifiable self-sovereign identities. In: 2020 IEEE 19th International Conference on Trust, Security and Privacy in Computing and Communications (TrustCom), pp. 1020–1027 (2020)
25. Author, F.: Article title. Journal **2**(5), 99–110 (2016)
26. Parameswarath, R.P., Gope, P., Sikdar, B.: User-empowered privacy-preserving authentication protocol for electric vehicle charging based on decentralized identity and verifiable credential. ACM Trans. Manage. Inf. Syst. (TMIS) **13**(4), 1–21 (2022)
27. Kortesniemi, Y., Lagutin, D., Elo, T., Fotiou, N.: Improving the privacy of IoT with decentralised identifiers (DIDs). J. Comput. Netw. Commun. (2019)
28. Bartolomeu, P.C., Vieira, E., Hosseini, S.M., Ferreira, J.: Self-sovereign identity: use-cases, technologies, and challenges for industrial IoT. In: 2019 24th IEEE International Conference on Emerging Technologies and Factory Automation (ETFA), pp. 1173–1180. IEEE (2019)
29. Fedrecheski, G., Rabaey, J.M., Costa, L.C., Ccori, P.C.C., Pereira, W.T., Zuffo, M.K.: Self-sovereign identity for IoT environments: a perspective. In: 2020 Global Internet of Things Summit (GIoTS), pp. 1–6, June 2020
30. Popov, S.: The tangle. White Pap. **1**(3), 30 (2018)
31. Niya, S.R., Jeffrey, B., Stiller, B.: KYoT: self-sovereign IoT identification with a physically unclonable function. In: 2020 IEEE 45th Conference on Local Computer Networks (LCN), pp. 485–490 (2020)
32. Weingaertner, T., Camenzind, O.: Identity of Things: applying concepts from self sovereign identity to IoT devices. J. Br. Blockchain Assoc. **4**(1), 1–7 (2021)

33. Pritikin, M., Richardson, M., Behringer, M., Bjarnason, S., Watsen, K.: Bootstrapping remote secure key infrastructures (BRSKI). Internet-Draft draft-ietf-anima-bootstrapping-keyinfra-18, IETF (2019)
34. Capela, F.: Self-sovereign identity for the Internet of Things: a case study on verifiable electric vehicle charging (Master's thesis) (2021)
35. Terzi, S., Savvaidis, C., Votis, K., Tzovaras, D., Stamelos, I.: Securing emission data of smart vehicles with blockchain and self-sovereign identities. In: 2020 IEEE International Conference on Blockchain, pp. 462–469 (2020)
36. Camenisch, J., Lysyanskaya, A.: A signature scheme with efficient protocols. In: Cimato, S., Persiano, G., Galdi, C. (eds.) Security in Communication Networks: Third International Conference, SCN 2002, Amalfi, Italy, 11–13 September 2002, Revised Papers 3, pp. 268–289. Springer, Cham (2003). https://doi.org/10.1007/3-540-36413-7_20
37. Gebresilassie, S.K., Rafferty, J., Morrow, P., Chen, L., Abu-Tair, M., Cui, Z.: Distributed, secure, self-sovereign identity for IoT devices. In: 2020 IEEE 6th World Forum on Internet of Things (WF-IoT), pp. 1–6 (2020)
38. Lamport, L., Shostak, R., Pease, M.: The Byzantine generals problem. In: Concurrency: The Works of Leslie Lamport, pp. 203–226 (1982)
39. Castro, M., Liskov, B.: Practical byzantine fault tolerance. In: OsDI, vol. 99, no. 1999, pp. 173–186 (1999)
40. Miller, A., Xia, Y., Croman, K., Shi, E., Song, D.: The honey badger of BFT protocols. In: Proceedings of the 2016 ACM SIGSAC Conference on Computer and Communications Security, pp. 31–42 (2016)
41. Cachin, C.: Architecture of the hyperledger blockchain fabric. In: Workshop on Distributed Cryptocurrencies and Consensus Ledgers (2016)
42. Luu, L., Narayanan, V., Zheng, C., Baweja, K., Gilbert, S., Saxena, P.: A secure sharding protocol for open blockchains. In: Proceedings of the 2016 ACM SIGSAC Conference on Computer and Communications Security, pp. 17–30. ACM (2016)
43. Kokoris-Kogias, E., Jovanovic, P., Gasser, L., Gailly, N., Ford, B.: OmniLedger: a secure, scale-out, decentralized ledger. IACR Cryptology ePrint Archive, vol. 2017, p. 406 (2017)
44. Zamani, M., Movahedi, M., Raykova, M.: RapidChain: scaling blockchain via full sharding. In: Proceedings of the 2018 ACM SIGSAC Conference on Computer and Communications Security, pp. 931–948 (2018)
45. Cong, K., Ren, Z., Pouwelse, J.: A blockchain consensus protocol with horizontal scalability. In: 2018 IFIP Networking Conference (IFIP Networking) and Workshops, pp. 1–9 (2018)
46. Sweeney, L.: Simple demographics often identify people uniquely. Health (San Francisco) 671(2000), 1–34 (2000)
47. Narayanan, A., Shmatikov, V.: Robust de-anonymization of large sparse datasets. In: 2008 IEEE Symposium on Security and Privacy, pp. 111–125 (2008)
48. Barbaro, M., Zeller, T., Hansell, S.: A face is exposed for AOL searcher no. 4417749. New York Times 9(2008), 8 (2006)
49. Ohm, P.: Broken promises of privacy: responding to the surprising failure of anonymization. UCLA l. Rev. 57, 1701 (2009)
50. How a Visa transaction works (2015). http://apps.usa.visa.com/merchants/become-a-merchant/ how-a-visa-transaction-works.jsp.
51. Aublin, P.L., Mokhtar, S.B., Quéma, V.: RBFT: redundant byzantine fault tolerance. In: 2013 IEEE 33rd International Conference on Distributed Computing Systems, pp. 297–306 (2013)

52. Singh, A., Das, T., Maniatis, P., Druschel, P., Roscoe, T.: BFT protocols under fire. In: NSDI, vol. 8, pp. 189–204 (2008)
53. Clement, A., Wong, E., Alvisi, L., Dahlin, M., Marchetti, M.: Making Byzantine fault tolerant systems tolerate Byzantine faults. In: Proceedings of the 6th USENIX Symposium on Networked Systems Design and Implementation. The USENIX Association (2009)
54. https://hyperledger.github.io/indy-did-method/#nym-transaction-version
55. Cristian, F., Aghili, H., Strong, R., Dolev, D.: Atomic broadcast: from simple message diffusion to Byzantine agreement. Inf. Comput. **118**(1), 158–179 (1995)
56. Bracha, G.: Asynchronous Byzantine agreement protocols. Inf. Comput. **75**(2), 130–143 (1987)
57. Cachin, C., Guerraoui, R., Rodrigues, L.: Introduction to Reliable and Secure Distributed Programming. Springer, Heidelberg (2011). https://doi.org/10.1007/978-3-642-15260-3

Biometric-Based Password Management

Pavlo Kolesnichenko[1] , Dmytro Progonov[2,3]([✉]) , Valentyna Cherniakova[2] ,
Andriy Oliynyk[4] , and Oleksandra Sokol[2]

[1] N-iX LTD, Kyiv, Ukraine
[2] Samsung R&D Institute Ukraine, Kyiv, Ukraine
{d.progonov,v.cherniakov}@samsung.com
[3] Igor Sikorsky Kyiv Polytechnic Institute, Kyiv, Ukraine
[4] Taras Shevchenko National University of Kyiv, Kyiv, Ukraine

Abstract. Major threat for the user's identity stem from selecting weak
passwords or re-using the same password for different systems. Modern
password managers are designed to address this human factor. But in
most cases this is achieved at cost of using a single master secret to
either derive access keys to protected services, or to encrypt a credentials database. Despite wide adoption, this boils down to security and
availability of this master secret.

We propose a technology to derive cryptographically-strong (of sufficient length and entropy) master secret from user's biometrics, such as
face and voice. If applied to password manager scenario, this allows to
amend or even completely replace master secret to avoid related risks.
While general approach (using fuzzy extractors) is known, the unique
part of the presented technology is small hint size (58KB for 128 bits
key) and low computational complexity (it takes 125 msec to extract the
key on Galaxy S22 phone in the worst case).

Experimental results show that FAR and FRR are close to 0% for wide
range of cryptographic keys lengths (from 80 to 256 bits). All computations are performed on-device, which means the technology is privacy-
friendly: user's biometrics never leaves the phone. The technology does
not require storing any sensitive data on the device, that is important
advantage in comparison with traditional biometric authentication solutions.

Keywords: Authorization · Password manager · Mobile devices

Acronym

BC	Biometric Cryptosystems
BKG	Biometric Key Generators
BTP	Biometric Template Protection
CK	Cryptographic Keys
DE-PAKE	Device-Enhanced Password Authenticated Key Exchange
DL	Digital Lockers

© The Author(s), under exclusive license to Springer Nature Switzerland AG 2023
R. Rios and J. Posegga (Eds.): STM 2023, LNCS 14336, pp. 23–41, 2023.
https://doi.org/10.1007/978-3-031-47198-8_2

FE	Fuzzy Extractors
FR	Facial Recognition
FAR	False Acceptance Rate
FRR	False Rejection Rate
OPRF	Oblivious Pseudo Random Function
PM	Password Manager

1 Introduction

Nearly every WEB service insists on creating a user account, which requires some form of authentication. Login/password is the most widespread form of authentication. To make it work, the user must select strong passwords [1,6,31] and avoid using the same password for distinct services [2,44,46]. Users rarely do this because of the risk to forget the password and lose access to the service. The industry has put significant efforts to build "passwordless future", including FIDO standards [8] based Single-Sign-On platforms and integrated identity management services, such as Google's One-Tap Android Sign-In [7]. The downside is: it requires a binding to a single identity provider, which ultimately ends up with remembering some master secret. Another approach is using specialized credentials management applications like Password Manager (PM). The PM is essentially an encrypted digital vault that stores the login information, credentials and sensitive data. In fact, modern PM do not solve the problem of remembering passwords but just relax it.

We propose a technology to produce a stable bit stream from user's biometrics, suitable for cryptographic operations (which means, of sufficient bit length and entropy). It relies on advanced methods for a Cryptographic Keys (CK) extraction and reproducing from fuzzy biometric data without storing any sensitive biometric templates on the device.

Practical application to credentials management solutions, such as password managers, allows to avoid master password management hurdles as well as diversification of access keys to distinct resources in a way that is natural and transparent for end user.

The key difference between modern methods for biometric-based user authentication and our proposed approach is an additional layer of protection for credentials. By using modern recognition technologies like FaceID, end-user obtains access to the data as soon as authentication is passed successfully. Our approach makes possible reliable restoration of encryption keys to decrypt already stored credentials without any inconveniences for the user. In addition, encryption keys can be generated for each service from a single biometric sample without decreasing security level. Regarding threat model, we consider active attacks on both biometric recognition and cryptographic components. We assume that underlying recognition algorithms include liveness detection in order to protect from various spoofing attacks.

The rest of this paper is organized as follows. Review of modern password managers for desktop and mobile devices can be found in Sect. 2. Proposed technology is described in Sect. 3. Section 3.1 presents the general description and security analysis for proposed scheme, while Sect. 3.2 provides description of used fuzzy extraction methods. Performance evaluation results are provided in Sect. 4 and are discussed in Sect. 5. Section 6 concludes the paper.

2   Related Works

The majority of modern WEB browsers offer at least a rudimentary built-in password managers[1]. These managers offer only basic features, such as automatic filling the login forms, passwords security checking and theirs storing in encrypted format. However, such PM are vulnerable to widespread tools for password extraction in browsers, like Mimikatz [19]. Therefore, specialized commercial applications like NordPass, LastPass and Bitwarden were proposed to effectively counteract such tools.

Despite rich functionality of advanced PM, they still require their users to remember a master secret. This approach has well-known security and usability implications, such as availability issues (if master password is forgotten) and security (if it is too simple or if the same password is re-used). Several approaches were proposed to make master secret management easier in the modern PM. One of them is to derive the whole passwords database from an initial master password, for example PwdHash [36]. Another one is to apply Device-Enhanced Password Authenticated Key Exchange (DE-PAKE) model [26]. Practical implementation of DE-PAKE model was proposed by Shirvanian et al. with Oblivious Pseudo Random Function (OPRF) scheme implemented in SPHINX [38]. The scheme transforms a human-memorable string (password) into a pseudo random one for each OPRF in password-agnostic way. This makes SPHINX robust to communication channel eavesdropping, dictionary and man-in-the-middle attacks [38]. Nevertheless, the selection of an appropriate OPRF for practical use still remains an unsolved task. Also, the SPHINX method preserves the necessity of using master password that should be remembered by the user.

The alternative approach for credential protection was proposed in iCloud Keychain by Apple [4]. The public and private keys pair are generated on each user device. The public key is shared with iCloud service, while private key never leaves the phone. Sharing credentials to a registered device is performed in several steps. First, sensitive data is encrypted N times using public keys of all devices that belong to the user. Then any of user's device becomes able to decrypt it, since its private key matches at least one of N public keys used for encryption.

Therefore, a secure and user-friendly (password-free) access to the password manager is a topical task. The paper is aimed at filling this gap by develop-

---

[1] For example, Firefox Monitor and Firefox Password Manager for Firefox, Google Password Manager for Chrome, Edge Password Manager for Microsoft Edge and Safari Password Manager for Safari browsers.

ing a biometric-based password-free credential data protection method to be integrated with PM.

3 Biometric-Based Password-Free Credentials Protection Method

Instead of using a master password, we propose to protect user's credentials by encrypting each service password with its unique CK derived from biometrics. This prevents master password to be a single point of failure and provides a handy way to manage credentials for many remote online resources without the need to select and memorize complicated passwords.

To make this approach work, a technology for generating crypto keys from raw biometric data is needed. In comparison with classic biometric authentication methods, this use case has essential security requirements: extracted bit streams must be stable up to a single bit, they must have enough entropy and have length sufficient to resist brute force attack. During analysis we considered threat model based on active attacks on both biometric recognition and cryptographic components. In addition, it is assumed that underlying recognition algorithms include liveness detection in order to protect from various spoofing attacks.

Practical implementation can be done using novel methods for Biometric Template Protection (BTP). These methods provide the following benefits that are important for PM [17]:

- Unlinkability: it is impossible to determine if two or more protected templates were derived from the same biometric instance, e.g. face. This property prevents cross-matching across different credentials databases.
- Irreversibility: it is impossible to reconstruct the original biometric data given a protected template and its corresponding auxiliary data. With this property fulfilled, the privacy of the users' data is increased. Also the security of the system is increased against presentation and replay attacks.
- Renewability (cancellability): it is possible to revoke old protected template and to create a new one from the same biometric instance and/or sample, e.g. face image. Thus it is possible to revoke and reissue the templates in case the database is compromised.
- Performance preservation: computational overhead caused by biometrics processing is acceptable for target applications.

The proposed methods for BTP can be divided into cancellable biometrics and Biometric Cryptosystems (BC) [35]. Cancellable biometrics employs transforms in signal or feature domains which enables a biometric comparison in the transformed (encrypted) domain [35]. In contrast, the majority of BC bind a sequence of bits (key) to a biometric feature vector resulting in a protected template [35]. That is, BC further allows for the derivation of digital keys from protected biometric templates, e.g. fuzzy commitment and fuzzy vault scheme.

Practical application of this approach requires the technology to produce strong (random) cryptographic keys that can be easily and quickly restored without bringing any inconveniences for the user. For this purpose, we propose an advanced BC, such as Fuzzy Extractors (FE), that are robust to natural input biometrics alterations and do not require on-device storage of sensitive data. Therefore, the PM will remain secured even if an attacker has physical access to user's mobile device, can extract credentials from one of user's device or used service.

Generation of CK requires the use of data source with enough entropy (at least 80 bits). The latter one can be used to construct a secure Biometric Key Generators (BKG) that provides key randomness and biometric privacy [12, 41,47]. Special interest was taken on security analysis of biometric sources of entropy, namely entropy of produced binary bit streams [12,28]. Recent advances in this area proved that the entropy level for widespread types of biometrics, like facial and iris images, is acceptable for security applications [18,45].

In the paper we consider the case of facial biometrics. Frontal camera is present in most of smartphones, which makes this type of biometrics a convenient option. Also a rich set of Facial Recognition (FR) methods to extract robust features with minimal processing overhead exists. A special type of BKG, namely FE construction, was proposed to mitigate variability of extracted facial features and to get stable CK even for noised samples. We assume that the underlying recognition algorithms include liveness detection in order to protect from spoofing attacks, such as copying or artificially reproducing biometric samples.

3.1 Procedure of Credential Data Protection

The proposed approach to protect user's credentials includes two procedures called "Password generation" and "Password retrieval". The processing pipeline for both procedures is presented in Fig. 1.

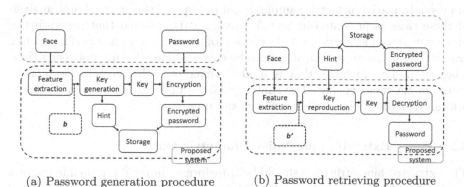

(a) Password generation procedure   (b) Password retrieving procedure

Fig. 1. Flowchart of password generation (a) and retrieving (b) procedures from inputted facial features of a user.

During generation procedure (Fig. 1a), credential data to be protected (encrypted) are taken. Then, the collected facial image of this user is fed into FR module to extract a feature vector **b**. Then the password and the feature vector **b** are passed to Gen algorithm of FE for calculation of encryption (secret) key **k** and public information (hint) **h**. Finally, password is encrypted using the generated key and then stored together with the corresponding hint.

The retrieving procedure (Fig. 1b) starts with getting a facial image of the user. Then, captured face image is fed into a FR module. Produced feature vector **b**′ and stored hint **h** are passed to Rep algorithm of FE to reproduce the secret key **k** and restore password. If a user is the same, i.e. vectors **b** and **b**′ are close enough, then the FE is able to reproduce the key **k** up to a single bit, so that the password can be decrypted.

The proposed approach assumes that biometric data is input to the device every time: storing any biometric template (for example, facial features) is not required. This allows to mitigate attacks based on re-using of stolen templates. Then, input facial images are processed by an independent module for face recognition (Fig. 1). In order to use biometric fuzzy data in FE this data must satisfy a couple of requirements [43]:

1. Features should be distributed almost uniformly in unitary cube or hypersphere, depending on the features of used recognition system.
2. The error rate for considered FE should be applicable for commercial usage in modern services. For example, the FAR and FRR should be less that $5 \cdot 10^{-4}$ and 10^{-1} respectively for FE to be deployed on mobile devices with Android OS [20].
3. The threshold t to separate features related to different users must be high enough to make the system secure in terms of robustness to spoofing. This can be presented as the dependency of the threshold on the features distribution as $(2t)^{-n}$ where n is the dimension of the features.

Therefore, the robustness of the face recognition module to altered, spoofed or even adversarial biometric samples is out of scope of this paper. Some analysis of these cases is presented in Sect. 5. Finally, it is assumed that processing of biometric as well as encryption/decryption of credentials data is performed in a trusted environment, so that it cannot be eavesdropped during processing time. The public data (a hint), that is used for encryption key restoration, can be stored in any memory unit. The Hint is public: it does not disclose any sensitive information about biometric samples of the user.

3.2   Fuzzy Extractors for Noisy Biometric Data

The extracted biometric data should be transformed into a reproducible uniform random bit stream (a crypto key **k**) that can be applied in cryptography-related tasks such as encryption and digital signing. We propose to apply special cryptographic constructions called fuzzy extractors [24] to address this task. The FE was designed to achieve information-theoretic security of information processing systems by converting noisy input (facial features in our case) to a uniformly distributed key.

The FE typically requires the use of generation (Gen) and reproduction (Rep) algorithms. The former one outputs a helper data h (a hint) and a generated key $\mathbf{k}$ on selected input biometric sample $\mathbf{b}$. Then, by given h and a new sample $\hat{\mathbf{b}}$ that is close to $\mathbf{b}$ as input, the Rep algorithm allows for recovering $\mathbf{k}$. The FE are secure in the sense that they do not reveal much about biometric data, i.e. h has no or little information about $\mathbf{b}$. This means that the hint h can be stored in a public storage without compromising the privacy of user's biometric data.

One of the state-of-the art scheme that we use is Digital Lockers (DL) fuzzy extractor [16]. The DL-scheme has reusability property, which means this construction remains secure even if a user enrolls multiple times with the same or correlated input data. We use the notation $\mathbf{c} = \mathsf{lock}(\mathbf{key}, \mathbf{val})$ for the algorithm that performs the locking of the value $\mathbf{val}$ using the key $\mathbf{key}$, and $\mathsf{unlock}(\mathbf{key}, \mathbf{c})$ for the unlocking algorithm. The unlocking algorithm outputs $\mathbf{val}$ if $\mathbf{key}$ is correct and empty set $\emptyset$ with high probability otherwise.

Let the source $B = \{b_i\}_{i=1}^n$ consist of n-elements string over some arbitrary alphabet $\mathcal{Z}$, in particular the binary alphabet $\{0, 1\}$. The generation algorithm for k-bits cryptographic key $\mathbf{k}$ over alphabet $\mathcal{Z}$ is represented in pseudo-code form in Algorithm 1 [16].

Data: biometric sample $\mathbf{b} \in \mathcal{Z}^n$
Result: key $\mathbf{k}$, set of digital lockers $\mathcal{P} = \{p_1, \ldots, p_l\}$
Sample $\mathbf{k} \leftarrow \{0, 1\}^k$;
Initialize empty set $\mathcal{P}$ of digital lockers ;
for $i = 1, \ldots, l$ do
    Choose uniformly random set $\mathcal{J}_i = \{j_m\}_{m=1}^k, j_m \in [1; n]$;
    Get sub-set of biometric elements: $\mathcal{V}_i = b_{j_{i,1}}, \ldots, b_{j_{i,k}}$;
    Set $c_i = \mathsf{lock}(\mathcal{V}_i, \mathbf{k})$;
    Append digital locker $p_i = \{c_i, \mathcal{J}_i\}$ to the set $\mathcal{P}$;
    Output key $\mathbf{k}$ and set $\mathcal{P}$.
end

Algorithm 1: Generation algorithm for Digital Lockers Fuzzy Extractor

---

At the first stage, a cryptographic key $\mathbf{k}$ is generated. Then, the random set of indices $\mathcal{J} = \{j_1, \ldots, j_k\}$ and the corresponding set of biometric elements $\mathcal{V} = \{b_{j_1}, \ldots, b_{j_k}\}$ are generated. At the second stage a set of digital lockers $\mathcal{P} = \{p_1, \ldots, p_l\}$ is being created. This set hides $\mathbf{k}$ using elements of subsets $\mathcal{V}$. The composition of several digital lockers allows us to re-use biometric samples. This makes it possible to revoke and regenerate CK without disclosing biometric.

The algorithm for reconstruction of cryptographic key $\mathbf{k}$ from new binary sample $\hat{\mathbf{b}}$ is presented in pseudo-code form in Algorithm 2 [16].

Data: new biometric sample $\hat{\mathbf{b}} \in \mathcal{Z}^n$, set $\mathcal{P} = \{p_1, \ldots, p_l\}$ of digital lockers
Result: restored key $\mathbf{k}' \in \{0,1\}^k$ or $\emptyset$
for $i = 1, \ldots, l$ **do**
    Parse digital locker p_i as c_i and $\mathcal{J}_i = (j_1, \ldots, j_k)$;
    Choose elements from biometric $\hat{\mathcal{V}}_i = \hat{b}_{j_1}, \ldots, \hat{b}_{j_k}$;
    Compute element of key $k_i = \mathsf{unlock}(\hat{\mathcal{V}}_i, c_i)$;
    if $k_i = \emptyset$ **then**
        | output empty set $\emptyset$;
    end
end
Output restored key $\mathbf{k}$.
Algorithm 2: Reproduction algorithm for Digital Lockers Fuzzy Extractor

Practical application of the DL extractor requires transformation of extracted facial features $\mathbf{b}$ into a vector with elements from some alphabet $\mathcal{Z}$ [16], for instance bit values. It can be achieved by processing biometric $\mathbf{b} = \{b_1, \ldots, b_n\}$ with a binarization function $Q(\cdot) : \mathbb{R}^n \to \{0,1\}^l$. The following types of binarization functions can be used for biometric sample $\mathbf{b}$ processing:

1. *Sign-based binarization*—bit values are extracted using the modified signum function:
$$Q_{sign}(b_i) = (\mathrm{sign}(b_i) + 1)/2, i \in [1; n]. \tag{1}$$

2. *Approximation-based binarization*—is closely related to well-known Shamir's secret sharing procedure [37]. Firstly, a set of elements is chosen using sliding window $\mathbf{w}$ of size w. Then, selected elements are used as reference points for extrapolating values outside of the window $\mathbf{w}$ with position $(w + \Delta), \Delta > 0$. Finally, bit value is extracted using extrapolated value's sign according to (1).

3. *Statistics-based binarization*—takes into account features of biometric's elements distribution, for example quantiles. At first, the disribution is split into parts using quantiles/percentiles as boundaries. Then, these parts are enumerated in ascending order. Binarization of new elements is achieved by taking binary representation of index for the nearest part to element.

We considered applying mentioned approaches for multidimensional vectors binarization. The approximation-based binarization was considered for two cases, namely for applying polynomial and trigonometric-based approximation methods. The former one was done by applying Lagrange multipliers method [15], while the latter one was based on solving following systems of linear equations:

$$\begin{cases} u_0 + \sum_{i=1}^{\lceil \hat{w} \rceil} u_i \cos\left(\frac{2\pi i}{K}\right) + \sum_{i=1}^{\lfloor \hat{w} \rfloor} u_{i+\lceil \hat{w} \rceil} \sin\left(\frac{2\pi i}{K}\right) &= b_1, \\ u_0 + \sum_{i=1}^{\lceil \hat{w} \rceil} u_i \cos\left(\frac{4\pi i}{K}\right) + \sum_{i=1}^{\lfloor \hat{w} \rfloor} u_{i+\lceil \hat{w} \rceil} \sin\left(\frac{4\pi i}{K}\right) &= b_2, \\ \dots &= \dots, \\ u_0 + \sum_{i=1}^{\lceil \hat{w} \rceil} u_i \cos\left(\frac{2\pi w i}{K}\right) + \sum_{i=1}^{\lfloor \hat{w} \rfloor} u_{i+\lceil \hat{w} \rceil} \sin\left(\frac{2\pi w i}{K}\right) &= b_w. \end{cases} \quad (2)$$

where $\hat{w} = (w - 1)/2$—be the middle element of the sliding window; K—period of trigonometric functions. The solution of the system (2) is a vector $\mathbf{u} = \{u_0, \dots, u_w\}$ used for calculation of extrapolated value:

$$b_{new} = x_0 + \sum_{i=1}^{\lceil \hat{w} \rceil} x_i \cos\left(\frac{2\pi i(w + \Delta)}{K}\right) + \sum_{i=1}^{\lfloor \hat{w} \rfloor} x_{i+\lceil \hat{w} \rceil} \sin\left(\frac{2\pi i(w + \Delta)}{K}\right).$$

Finally, the bit value is extracted by applying sign-based binarization function (1) to extrapolated value.

The statistics-based binarization was done using Gray codes [21,22]. First, the interval of biometric sample's $\mathbf{b}$ elements was divided into S parts using either predefined boundaries q_s, or percentiles pre-calculated from elements distribution. These parts were indexed in increasing order from 0 to $(S - 1)$. Then, binary representations of indices were transformed into Gray code. Finally, new element of biometric was binarized by taking the corresponding Gray code of the nearest part to the element.

For the sake of brevity, we will call sign-based binarization function (1) as sign-based binarizer, while approximation-based methods will be denoted as polynomial and trigonometric binarizers. The statistics-based binarization method will be denoted as Gray-code binarizer.

Also, we propose using "stable" bits of biometric's binarized elements during sampling stage in Gen algorithm of DL-extractor. The idea behind this proposal is based on selection positions of bits in binarized features that remain the same for new estimations. The stability of ith bit from binarized feature of user was estimated according to the following formula:

$$s_i^j = \Pr(b_i = 0), i \in [1; l], j \in \{0, 1\}, \quad (3)$$

where $\mathbf{s}^0 = (s_1^0, \dots, s_l^0)$ and $\mathbf{s}^1 = (s_1^1, \dots, s_l^1)$ are vectors of probability to obtain either "0", or "1" bit values. Probabilities $\Pr(b_i = 0)$ and $\Pr(b_i = 1)$ can be estimated using the set of user's binarized features. Finally, elements of vectors $\mathbf{s}^0$ and $\mathbf{s}^1$ are thresholded to determine the indices (positions) of elements that have high probability to be binarized into "0" or "1". The accuracy of proposed reproduction algorithms is discussed in Sect. 4.

4 Experiments

Performance evaluation of proposed approach was done using FERET [33,34] and VGG [32] datasets. The evaluation process was divided into several steps.

First, the images from the datasets were processed by the standard OpenCV library [14] to detect and crop users faces.

At the second stage, facial features were extracted from prepared images using state-of-the-art VGG-Face [32] and Sphere-Face [29] networks. The output fully-connected layer of pre-trained networks was removed to obtain the estimated features. Thus, the feature's dimensionality was 4,096 for VGG-Face and 512 for Sphere-Face networks.

Then, the cross-feature distances were estimated for intra-user (related to the same user) and inter-user (across different users) case. Estimated intra-user and inter-user features distributions for VGG-Face and Sphere-Face networks on considered datasets are shown in Fig. 2.

Features related to the same user are compactly packed, whereas features related to different user are located apart for SphereFace extractor (Fig. 2c). On the other hand, inter-user and intra-user features distribution are partially overlapped by feature extraction from FERET dataset (Fig. 2d). In contrast, the corresponding distributions are highly overlapped for both VGG (Fig. 2a) and FERET (Fig. 2b) datasets if VGG-Face network is used. Thus, the distances between the features of the same user and other users are relatively small, which may negatively impact FE performance.

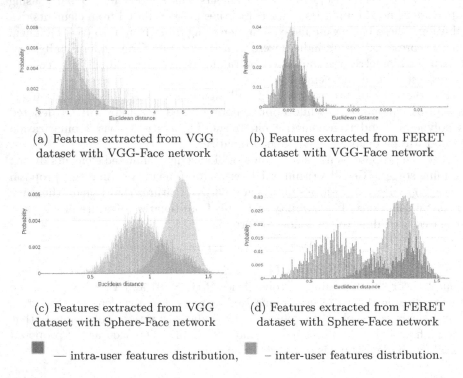

(a) Features extracted from VGG dataset with VGG-Face network

(b) Features extracted from FERET dataset with VGG-Face network

(c) Features extracted from VGG dataset with Sphere-Face network

(d) Features extracted from FERET dataset with Sphere-Face network

— intra-user features distribution,   – inter-user features distribution.

Fig. 2. Histogram of cross-features distances for the same (intra-user) or different (inter-user) users features distribution, extracted by VGGFace (a–b) and SphereFace (c–d) networks.

On the next stage, extracted features were binarized using sign-based Gray-code polynomial and trigonometric binarizers with 4-elements sliding window. Gray-code binarizer used 8 quantization boundaries q_s, and the shift Δ was equal to 0.5 for polynomial and trigonometrical binarizers. Binarized features were given to DL-fuzzy extractor for CK generation. The threshold value $T = 0.96$ was used for selection of "stable" bits (3) from binarized features. The lengths of obtained keys were: 80, 96, 128, 160, 192 and 256 bits.

Finally, the FAR and FRR were estimated using CK generated by DL-extractor. The estimations were performed by applying the standard cross-validation approach—the features for each user were divided into training (70%) and test (30%) subsets. Then, the CK and corresponding helper data (set $\mathcal{P}$ of digital lockers) were generated for each subset. The FAR was estimated using the set $\mathcal{P}_i$ of the ith user to reproduce the corresponding key $\mathbf{k}_i$ by $j^{th}(i \neq j)$ user's features. The FRR was estimated by comparison of keys generated for i^{th} user on training part ($\mathbf{k}_i^{train}$) and reproduced on the test part ($\mathbf{k}_i^{test}$). Cross-validation procedure was performed 100 times to estimate the averaged values of FAR and FRR.

The case of usage a of a small number of locks ($N_L = 5$) was considered. The box plots for FAR and FRR values estimated for the case of VGG-Face an Sphere-Face networks are presented in Figs. 3 and 4.

Estimated values of FRR are close to 0% for VGG-Face network for all considered binarizers (Fig. 3c–Fig. 3d). Also, the FAR values are decreasing if longer CK are used for both considered datasets (Fig. 3a-Fig. 3b). Nevertheless, the mean values (middle line within box plots) and the variation (heights of box plots) of FAR remain high (for more than 10%) for sign-based and polynomial binarizers. On the other hand, the use of Gray code and polynomial binarizers allows for preserving low (close to 0%) values of FAR, while obtaining outliers (FAR is close to 100%) for some users.

Note that increasing N_L value during key $\mathbf{k}$ restoration would not lead to changes of FRR values, since we have obtained close to 0% values even for $N_L = 5$ locks. On the other hand, getting more trials (locks) may lead to the corresponding increase of FAR values. Therefore, the use of VGG-Face network based on FR module (Fig. 3) is not suitable for real-life applications.

The results obtained for Sphere-Face network (Fig. 4) drastically differ from the previous case (Fig. 3). Here, we obtained close to 0% FAR (Fig. 4a–Fig. 4b), but FRR has grown (Fig. 4c–Fig. 4d). Note that FRR values are considerably less for the VGG dataset (zero mean with up to 3% variation, Fig. 4c) in comparison with FERET dataset (close to 100% mean with up to 25% variation, Fig. 4d). It can be explained by the differences of images collection methodology—samples from FERET dataset were collected in controlled environment, while VGG dataset comprises of images captured in various conditions (for various illumination, users age, presence of makeup). Thus, the features obtained from VGG dataset are much more diverse than those from FERET. It leads to more accurate selection of "stable" bits for binarized vectors and, correspondingly, the increase of robustness of proposed method to variations of user's facial images.

34   P. Kolesnichenko et al.

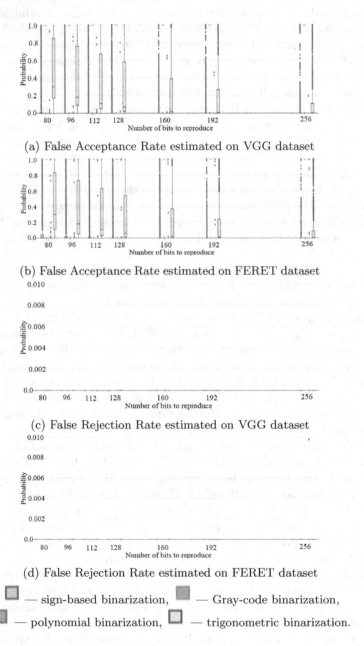

(a) False Acceptance Rate estimated on VGG dataset

(b) False Acceptance Rate estimated on FERET dataset

(c) False Rejection Rate estimated on VGG dataset

(d) False Rejection Rate estimated on FERET dataset

— sign-based binarization,   — Gray-code binarization,

— polynomial binarization,   — trigonometric binarization.

Fig. 3. Box plots of False Acceptance Rate (a–b) and False Rejection Rate (c–d) depend on cryptographic keys size for the users facial features extracted with VGG-Face network.

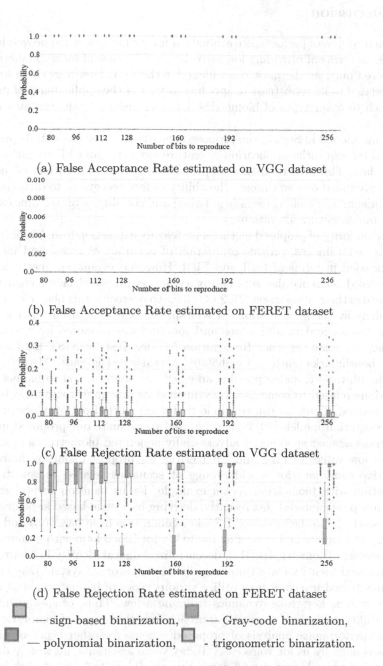

(a) False Acceptance Rate estimated on VGG dataset

(b) False Acceptance Rate estimated on FERET dataset

(c) False Rejection Rate estimated on VGG dataset

(d) False Rejection Rate estimated on FERET dataset

— sign-based binarization,   — Gray-code binarization,

— polynomial binarization,   - trigonometric binarization.

Fig. 4. Box plots of False Acceptance Rate (a–b) and False Rejection Rate (c–d) depend on cryptographic keys size for the users facial features extracted with Sphere-Face network.

5 Discussion

Obtained results of performance evaluation for proposed scheme proved its effectiveness in terms of providing low error levels. The case of facial biometrics was considered only, as the most convenient for the user. However, there are questions related to its robustness to spoofing as well as the applicability of proposed approach to other types of biometrics. Let us consider these questions in more details.

As mentioned in Sect. 3, the proposed construction for credentials protection is based on sequential application of feature extractor and FE to gathered biometric data. Thus, the robustness of proposed scheme to alterations of biometric data is grounded on two things—the ability of face recognizer to distinguish features for similarly looking users (e.g. twins) and stability of bit streams extracted by FE under feature alterations.

The majority of proposed methods for feature extraction from biometric data, such as facial images, provide evaluation of accuracy of cross- and inter-users identification in terms of FAR and FRR. However, evaluation methodology as well as used reference datasets may vary across proposed face recognizer that complicates their comparison [27, 29, 30, 32]. To overcome this obstacle, a number of publicly available evaluations of widespread face recognition models robustness to biometric data alterations and spoofing was proposed by governmental agencies (for example, Face Recognition Vendor Test by NIST [23]) and open source benchmarks (such as Celeb500K, MegaFace, DigiFace-1M [11]). On the other hand, the FE makes possible an effective counteraction to biometric data alterations related to some user only instead of providing robustness to spoofing in general. Thus, we can conclude that the reported performance of applied face recognition models [29, 32] can be used to evaluate our proposed method's robustness against spoofing or adversarially generated biometric samples.

The low values of FAR and FRR errors for proposed solution makes it an attractive candidate for a wide range of scenarios related to sensitive data restoration with biometrics. As an example, let us mention several examples that drew public interest just recently: deriving biometric-based keys for a crypto wallet seed [3], two-factor-based RSA key generation from password and biometrics [40], to generate and recover a private key for Blockchain with biometrics [10], Self-Sovereign Biometric IDs [13], decentralized digital identity in Metaverse [5], decentralized Identity Data Hub with distributed storage system [25], etc. However, mentioned use cases may utilize various types of biometrics, such as facial images, voices, keystroke dynamics to name a few. Thus, of special interest is applicability analysis of proposed approach for other types of biometric data.

The performance analysis of proposed scheme for other types of biometric data was done by us for voice case. Today voice assistants are widely deployed. This makes voice biometrics an attractive candidate to be used with proposed method for user's credential data protection. In the paper, we considered the case of speech data obtained from the standard Mozilla Common Voice open dataset [9]. The sub-set of 51,000 speakers with at least three audio files for each speaker was used during our evaluation.

The speech signals were processed in several steps. First, the modern X-VectorsNet neural network [39] was applied to extract features from the speech. Then, obtained features were binarized using Gray code similar to Sect. 3.2. Finally, the binarized features were processed usingDL extractor. The locker's size was varied from 66 to 84 bits with the step of 2 bits, while the amount of lockers varied from 9, 000 to 12, 000 with the step 1, 000.

The dependencies of FAR and FRR on parameters of used DL fuzzy extractor estimated on Mozilla Common Voice dataset are presented in Fig. 5.

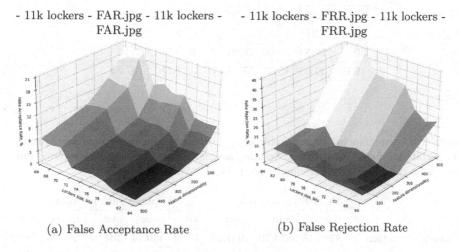

(a) False Acceptance Rate   (b) False Rejection Rate

Fig. 5. The dependencies of FAR (a) and FRR (b) on the extracted acoustic features dimensionality and bitsize of lockers at fixed number (11,000) of lockers for DL fuzzy extractor. The standard deviation of estimated error level for other number of lockers does not exceed 1% for FAR and 2% for FRR respectively. Note that values at X and Y axes for (a–b) are sorted in reverse order for better clarity of presented plots.

The estimated FAR and FRR values are similar for both voice (Fig. 5) and facial biometrics (Fig. 3 and 4)—close to 0% FAR and near 5% FRR. This proved the effectiveness of proposed approach in case of applying the technology to different types of biometrics. Obtained results for voice biometric conformed with the evaluation reports for the state-of-the-art commercial systems for speaker identification [42]. Nevertheless, note that the performance of the approach considerably depends on the feature extractor that was used. This means feature extractor must be selected carefully.

For practical application on mobile devices, computational efficiency and low memory utilization are important. Proposed method requires to store a piece of public data (a hint) and to process the inputted facial images by neural networks. The evaluation of computational and storage overhead was done using the prototype for smartphone Samsung Galaxy S22 (with Android 13 on board). Measured memory consumption and the duration of CK generation/retrieving procedures are represented in Table 1 and Table 2 respectively.

Note that these evaluation results were obtained for CPU-only execution of considered neural networks. The processing duration can be reduced if mobile device's graphical or neural hardware processing units are used.

Table 1. Dependency of hint size on cryptographic key size and locks number. Measurements are presented in KB.

Locks	Cryptographic key size (bits)					
	80	96	128	160	192	256
5	1.95	2.27	2.89	3.52	4.14	5.39
10	3.91	4.53	5.78	7.03	8.28	10.78
50	19.53	22.66	28.91	35.16	41.41	53.91
100	39.06	45.31	57.81	70.31	82.81	107.81

Table 2. Dependency of cryptographic keys generation/retrieving procedure duration for 128 bits key for VGG-Face and Sphere-Face networks. The case of sign-based binarizer (1) is considered. Measurements are presented in milliseconds.

		VGG-Face network				Sphere-Face network			
		max	min	mean	std	max	min	mean	std
Feature extraction		1215.65	854.59	1081.28	45.28	466.74	304.92	365.45	28.92
Key generation		144.12	114.28	124.19	8.97	77.37	37.42	56.99	12.99
Key retrieval	success	68.77	1.52	17.45	19.44	15.17	1.52	11.93	6.34
	fail	137.54	114.28	125.16	7.49	65.74	48.55	56.08	5.04

The hint size scales linearly with the number of locks. It remains relatively small (up to 58 KB) even for 128-bits cryptographic key and 100 locks for used fuzzy extractor (Table 1).

The mean time of features extraction procedure differs for VGG-Face and Sphere-Face networks up to 3 times (Table 2). The reason is significant difference of layers count—38 layers for VGG-Face and only 10 layers for Sphere-Face networks. The mean duration of key generation is relatively small for both networks (about 124 ms for VGG-Face and 57 ms for Sphere-Face) and depends only on parameters of DL-extractor. On the other hand, the key retrieval duration may vary substantially—from 11.9 ms (only part of locks are used for successful key recovery) to 125.2 ms (all locks are used, while key **k** was not restored).

6   Conclusion

We proposed a technology to produce a cryptographic key with essential security properties (entropy, length and bit stream stability) from user's biometrics.

Unique features are small hint size and low computational overhead. The reliable extraction of keys from noised biometric samples (facial images and speech signals) is achieved by using DL fuzzy extractor.

The prototype of proposed system was developed and tested on the standard VGG and FERET datasets of facial images and Samsung Galaxy S22 phone. Evaluation results confirmed close-to-zero values of FAR and FRR metrics (near 0%) in case of real-life images (VGG dataset) and applying of state-of-art Sphere-Face convolutional neural network for feature extraction. In addition, the advantage of proposed approach is small size of public information to be used for CK restoration—about 58 KB hint size for 128-bit key. The important feature of proposed technology is the ability to use various types of biometrics as the sources for encryption key. In this work we used facial images and voice. However, others types of biometrics, such as fingerprint and iris, can also be considered.

The technology can be applied to improve modern password managers. It is privacy-friendly (sensitive biometric data never leaves the device), secure against device loss (no sensitive data is ever stored on the device) and does not require the user to select and remember any master password. In comparison with the existing state-of-the-art methods, such as FaceID, proposed approach provides the way to use biometrics not only for user identification, but also for cryptographic scenarios (encryption and digital signing).

References

1. Most hacked passwords revealed as UK cyber survey exposes gaps in online security. National Cyber Secyurity Centre (2019). https://www.ncsc.gov.uk/news/most-hacked-passwords-revealed-as-uk-cyber-survey-exposes-gaps-in-online-security
2. 2020 end-of-year data breach report. Technical report, Identity Theft Resource Center (2020). https://www.idtheftcenter.org/data-breaches/
3. Card-based crypto hardware wallet: Protecting crypto wallet private keys and transactions with a biometric card. Idemia Inc. report (2022). https://www.idemia.com/card-based-crypto-hardware-wallet
4. Set up iCloud Keychain. Apple Inc. (2022). https://support.apple.com/en-gb/HT204085
5. The role of biometrics in the metaverse. CoinTelegraph Inc. report (2022). https://cointelegraph.com/metaverse-for-beginners/the-role-of-biometrics-in-the-metaverse
6. Top 200 most common passwords. NordPass Inc. (2022). https://nordpass.com/most-common-passwords-list/
7. Overview of One Tap sign-in on Android. Google Inc. (2023). https://developers.google.com/identity/one-tap/android/overview
8. User Authentication Specifications Overview. FIDO Alliance (2023). https://fidoalliance.org/specifications/
9. Ardila, R., et al.: Common voice: a massively-multilingual speech corpus (2019). https://doi.org/10.48550/ARXIV.1912.06670
10. Aydar, M., Cetin, S.C., Ayvaz, S., Aygun, B.: Private key encryption and recovery in blockchain (2019). https://doi.org/10.48550/ARXIV.1907.04156
11. Bae, G., et al.: DigiFace-1M: 1 million digital face images for face recognition (2022). https://doi.org/10.48550/ARXIV.2210.02579

12. Ballard, L., Kamara, S., Reiter, M.: The practical subtleties of biometric key generation. In: 17th USENIX Security Symposium (2008)
13. Bathen, L.A.D., et al.: Selfis: self-sovereign biometric IDs. In: 2019 IEEE/CVF Conference on Computer Vision and Pattern Recognition Workshops (CVPRW), pp. 2847–2856 (2019)
14. Bradski, G.: The OpenCV library. Dr. Dobb's J. Softw. Tools (2000)
15. Bramanti, M.: Matematica: Calcolo Infinitesimale e Algebra Lineare. Zanichelli, Bologna (2004)
16. Canetti, R., Fuller, B., Paneth, O., Reyzin, L., Smith, A.: Reusable fuzzy extractors for low-entropy distributions. Technical report, Cryptology ePrint Archive (2017). https://eprint.iacr.org/2014/243.pdf
17. Inernational Technical Committee: ISO/IEC 24745:2011. Information technology - Security techniques - Biometric information protection. Technical report, International Organization for Standardization and International Electrotechnical Committee (2011). https://www.iso.org/standard/52946.html
18. Daugman, J.: Information theory and the IrisCode. IEEE Trans. Inf. Forensics Secur. **11**, 400–409 (2015)
19. Delpy, B., Le Toux, V.: mimikatz. GitHub repository (2020). https://github.com/ParrotSec/mimikatz
20. Google: Measuring Biomentric Unlock Security (2020). https://source.android.com/security/biometric/measure
21. Gray, R., Neuhoff, D.: Quantization. IEEE Trans. Inf. Theory **IT-44**(6), 2325–2383 (1998)
22. de Groot, J., Škorić, B., de Vreede, N., Linnartz, J.-P.: Quantization in zero leakage helper data schemes. EURASIP J. Adv. Sig. Process. **2016**(1), 1–13 (2016). https://doi.org/10.1186/s13634-016-0353-z
23. Grother, P., Ngan, M., Hanaoka, K., Yang, J.C., Hom, A.: FRVT 1:1 verification. Technical report, National Institute of Standards and Technology (2022). https://pages.nist.gov/frvt/html/frvt11.html
24. Herder, C., Ren, L., van Dijk, M., Mandel Yu, M., Devadas, S.: Trapdoor computational fuzzy extractors and stateless cryptographically-secure physical unclonable functions. IEEE Trans. Depend. Secure Comput. **14**, 65–82 (2017)
25. Hersey, F.: Iris biometrics integrated with DIDH for 'most secured' data system for blockchain, metaverse. BiometricUpdate Site, News (2022). https://www.biometricupdate.com/202207/iris-biometrics-integrated-with-didh-for-most-secured-data-system-for-blockchain-metaverse
26. Jarecki, S., Krawczyk, H., Shirvanian, M., Saxena, N.: Device-enhanced password protocols with optimal online-offline protection. In: ACM Asia Conference on Computer and Communications Security (ASIACCS 2016). ACM (2016)
27. Kim, I., et al.: DiscFace: minimum discrepancy learning for deep face recognition. In: Ishikawa, H., Liu, C.-L., Pajdla, T., Shi, J. (eds.) ACCV 2020. LNCS, vol. 12626, pp. 358–374. Springer, Cham (2021). https://doi.org/10.1007/978-3-030-69541-5_22
28. Lim, M.H., Yuen, P.: Entropy measurement for biometric verification systems. IEEE Trans. Cybern. **46**, 1065–1077 (2015)
29. Liu, W., Wen, Y., Yu, Z., Li, M., Raj, B., Song, L.: SphereFace: deep hypersphere embedding for face recognition. In: The IEEE Conference on Computer Vision and Pattern Recognition (CVPR 2017) (2017)
30. Meng, Q., Zhao, S., Huang, Z., Zhou, F.: MagFace: a universal representation for face recognition and quality assessment (2021). https://doi.org/10.48550/ARXIV.2103.06627

31. Miessler, D., Haddix, J.: SecList: the Pentester's companion. GitHub repository (2022). https://github.com/danielmiessler/SecLists
32. Parkhi, O.M., Vedaldi, A., Zisserman, A.: Deep face recognition. In: British Machine Vision Conference (2015)
33. Phillips, P., Moon, H., Rizvi, S., Rauss, P.: The FERET evaluation methodology for face recognition algorithms. IEEE Trans. Pattern Anal. Mach. Intell. **22**, 1090–1104 (2000)
34. Phillips, P., Wechsler, H., Huang, J., Rauss, P.: The FERET database and evaluation procedure for face recognition algorithms. Image Vis. Comput. **16**(5), 295–306 (1998)
35. Rathgeb, C., Merkle, J., Scholz, J., Tams, B., Nesterowicz, V.: Deep face fuzzy vault: implementation and performance, November 2021
36. Ross, B., Jackson, C., Miyake, N., Boneh, D., Mitchell, J.: Stronger password authentication using browser extensions. In: USENIX Security Symposium (USENIX 2005) (2005)
37. Shamir, A.: How to share a secret. Commun. ACM **22** (1979)
38. Shirvanian, M., Jareckiy, S., Krawczykz, H., Saxena, N.: SPHINX: a password store that perfectly hides passwords from itself. In: IEEE 37th International Conference on Distributed Computing Systems (ICDCS 2017). IEEE (2017)
39. Snyder, D., Garcia-Romero, D., Sell, G., Povey, D., Khudanpur, S.: X-Vectors: robust DNN embeddings for speaker recognition. In: 2018 IEEE International Conference on Acoustics, Speech and Signal Processing (ICASSP), pp. 5329–5333 (2018). https://doi.org/10.1109/ICASSP.2018.8461375
40. Suresh, K., Pal, R., Balasundaram, S.R.: Two-factor-based RSA key generation from fingerprint biometrics and password for secure communication, **8**, 3247–3261 (2022). https://doi.org/10.1007/s40747-022-00663-3
41. Tambay, A.A.: Testing fuzzy extractors for face biometrics: generating deep datasets. Master's thesis, University of Ottawa, Ottawa, Canada (2020). https://doi.org/10.20381/ruor-25653
42. Team, S.D.: Personalized Hey Siri. Technical report, Apple Inc. (2018). https://machinelearning.apple.com/research/personalized-hey-siri
43. Tian, Y., Li, Y., Deng, R.H., Sengupta, B., Yang, G.: Lattice-Based Remote User Authentication from Reusable Fuzzy Signature. IACR Cryptology ePrint Archive 2019, 743 (2019)
44. Toubba, K.: Notice of recent security incident in 2022 year. Technical report, Last-Pass Inc. (2022). https://blog.lastpass.com/2022/12/notice-of-recent-security-incident/
45. Wang, Y., Yang, C., Shark, L.K.: Method for estimating potential recognition capacity of texture-based biometrics. IET Biometrics **7**, 581–588 (2018)
46. Whittaker, Z.: Norton LifeLock says thousands of customer accounts breached. Technical report, TechCrunch Inc. (2023). https://techcrunch.com/2023/01/15/norton-lifelock-password-manager-data/
47. Zhang, K., Cui, H., Yu, Y.: Facial template protection via lattice-based fuzzy extractors. Cryptology ePrint Archive, Paper 2021/1559 (2021). https://eprint.iacr.org/2021/1559

'State of the Union': Evaluating Open Source Zero Trust Components

Tobias Hilbig[1]($\boxtimes$)(iD), Thomas Schreck[1](iD), and Tobias Limmer[2](iD)

[1] HM Munich University of Applied Sciences, Munich, Germany
{tobias.hilbig,thomas.schreck}@hm.edu
[2] Siemens AG, Munich, Germany
tobias.limmer@siemens.com

Abstract. Zero Trust Architecture (ZTA) is a security model based on the principle "never trust, always verify". In such a system, trust must be established for both the user and the device for access to be granted. While industry adoption of commercial ZTA solutions is accelerating, the state of open-source implementations has yet to be explored. To that end, we survey open-source implementations of zero trust components and put forward a set of ZTA specific requirements to evaluate against. We also identify seven major challenges that hinder the adoption and deployment of open-source zero trust solutions. Our results show that implementations for individual components are much more mature compared to "all-in-one" ZTA solutions. The interoperability between solutions and the development of inter-component protocols are the main areas in which improvements can be made. Despite encouraging developments, we conclude that building ZTAs on top of open-source components is difficult.

Keywords: zero trust architecture · distributed systems security · authentication · authorization

1 Introduction

Network security has always been adapted and improved, both operationally and conceptually, to cope with new requirements and challenges of an ever-changing IT landscape. Despite this, the volume and impact of attacks against IT systems are increasing yearly. Existing perimeter-based network security cannot cope with today's requirements: Due to work-from-home, cloud computing, and BYOD policies, an organization's perimeter can no longer be clearly defined and protected. Furthermore, as the software and hardware landscape is becoming more diverse, enforcing strict access control policies in such heterogeneous environments can be challenging.

Zero Trust Architecture (ZTA) is one of the more recent concepts in network security. It can be summarized as "never trust, always verify" [1]. In contrast to the existing perimeter-based network security model, no inherent trust or privileges are granted based on the user's or device's physical or logical location.

R. Rios and J. Posegga (Eds.): STM 2023, LNCS 14336, pp. 42–61, 2023.
https://doi.org/10.1007/978-3-031-47198-8_3

Instead, a trust algorithm mediates access based on user, device, and service authentication and authorization. ZTA is a promising, significant change in network security philosophy compared with existing approaches and ideas.

While ZTA was conceived almost 20 years ago [2], it took considerable time for the first large-scale implementations. After a significant data breach known as "Operation Aurora" in 2009, Google implemented ZTA and subsequently published an article series [3] called "BeyondCorp" in 2014. Netflix [4] and Microsoft [5] also adopted ZTA in the last few years. Market research [6] shows that large parts of the industry are planning to transition their networks to ZTA in the near to medium-term future. To that end, "ZTA-as-a-service" offers by large enterprises seem to be the primary driving factor. It appears that the availability and maturity of open-source software and protocols necessary for implementing ZTA is an under-researched field.

Therefore, we want to answer the following research questions: (1) What are specific requirements for ZTA components? (2) To what extent do existing and emerging ZTA software solutions and protocols meet these requirements?

Our contributions are: (1) an overview of currently available and emerging ZTA software and protocols, (2) an evaluation of their features and maturity and (3) an analysis of specific requirements for ZTA components. In addition, our results can be used to select suitable solutions for a specific context.

The remainder of this work is structured as follows: We begin by discussing related work in Sect. 2 and lay out our methodology in Sect. 3. ZTA components and protocols, together with their requirements, are defined and analyzed in Sect. 4. Existing and emerging ZTA software and protocols are discussed in Sect. 5. We evaluate the solutions and present our results in Sect. 6. Challenges and future work are discussed in Sect. 7 and the paper concludes with Sect. 8.

2   Related Work

High-level requirements, or principles, for ZTA are well established and have been discussed extensively in the literature. NIST [1] defines seven tenets for ZTA. Kindervag, in [7], laid out reasons and key concepts for ZTA. ZTA core principles from a business perspective and example use-cases were proposed in a white paper by The Open Group [8]. In a later publication from the same consortium, these principles form the basis for nine high-level commandments for ZTA. While a universally accepted definition of ZTA is yet to be established, consensus for all high-level requirements formulated in these publications emerged.

Requirements have also been discussed for specific use-cases: Rose, in [9], proposes requirements for the implementation of ZTA at federal agencies. Rules for integrating legacy devices into Industrial Control Systems (ICS) based on ZTA were proposed by Køien, see [10]. The literature also provides requirements, challenges and lessons learned regarding real-world ZTA realization. Google published an article series dubbed "BeyondCorp" in 2014, see [3], discussing their approach and application of ZTA in great detail. Similarly, Netflix presented their implementation of ZTA in 2018, see [4]. While specific applications or use-cases can have more nuanced demands, the general principles for ZTA still hold.

For this reason, our aim is to formulate and evaluate requirements in an agnostic manner.

Three major surveys of the literature concerning ZTA have been published in the last two years. Yuanhang et al., see [11], conducted a survey of the academic literature, analyzing and comparing ZTA, identity and authentication, access control mechanism, and trust evaluation mechanisms. This work focused on advantages and disadvantages, current challenges and future research trends for ZTA. Buck et al., see [12], did a systematic literature review including gray literature. The authors state that ZTA has been gaining more and more interest in both academia and practice over the last few years, as measured by the number of publications per year. They also state that the literature primarily focuses on conceptual issues and benefits of ZTA, while user-related aspects and possible drawbacks are neglected. Finally, Seyed et al., see [13], surveyed the literature and specifically discussed authentication mechanisms, access control schemes, and encryption in the context of ZTA.

We also reviewed academic work with regard to federation in ZTA, as this concept is highly relevant for ZTA components. The topic was explored by Olson et al., see [14]. In this work, four design objectives for a distributed trust mechanism were given. These design objectives are further broken down into requirements resulting in a federated zero trust architecture in which trust is established via an additional, external proxy component. The concept of Zero Trust Federation (ZTF), together with a proof of concept implementation, was published by Hatakeyama et al., see [15]. Federated operation raises privacy concerns as context information about the user must be exchanged. The ZTF approach allows users to retain control over this information when exchanging with third parties. This is made possible through the use of the Continuous Access Evaluation Protocol (CAEP) [16] and User Managed Access (UMA) [17], an extension for OAuth2.0 [18].

While some publications in the area of zero trust touch the question of requirements for specific components, to the best of our knowledge this is the first academic work that collects and summarizes requirements for each component of a ZTA and evaluates available software solutions and protocols against these requirements.

3   Methodology

All data used in this work was acquired in April 2023. Google Scholar, the BASE database and the backward snowballing technique [19] were employed to collate relevant academic publications. While we did not use a specific search string, we included works in the field of zero trust that discuss general and specific requirements, broad surveys, and federation related publications in the context of zero trust architecture. Gray literature and inaccessible documents were excluded from the results.

Google, Github and the backward snowballing technique were used to find suitable open-source zero trust software solutions and protocols. Our keywords for this search were *zero trust software, zero trust implementation* and *open source zero trust*. We excluded abandoned, undocumented and proprietary solutions.

The selection of requirements for components was done as follows: Requirements concerning components were collected from related work, the primary source being the NIST standard. We extended this initial set by adding needed requirements for inter-component communication and federated operation. The resulting set of requirements was then reviewed and discussed individually with each member of our group. This review process was repeated twice, at which point consensus emerged.

4 Architecture and Requirements

For this work, we use NIST's definition of ZTA: "Zero trust architecture (ZTA) is an enterprise's cybersecurity plan that utilizes zero trust concepts and encompasses component relationships, workflow planning, and access policies. [...]" [1]. Moreover, we note that NIST discusses four primary components: The Policy Enforcement Point, the Policy Decision Point, the Policy Information Point, and optionally a client-side agent. These components were initially defined in XACML [20]. They are used to establish trust and mediate access to resources by evaluating authentication, authorization and assurance information for users, devices and services.

Before discussing the components and defining their specific requirements, we reflect upon the general architecture of ZTA and the vital concept of federation between ZTAs. In addition, we discuss the "control plane", i.e., the inter-component communication layer. All requirements listed in the following sections, together with a detailed description, can be found in Appendix A.

4.1 Architecture

ZTA is a fundamentally different approach to network security compared to existing perimeter-based networking. Instead of defining multiple zones with different "trust levels" such as the Internet, a demilitarized zone (DMZ) and the intranet, in ZTA, every asset and the network between assets is considered untrusted by default. In perimeter-based networking, much effort is made to separate these zones, for example by using firewalls and employing physical access control. While ZTA does not attempt to separate assets into zones, a strict separation is done based on the content of transmitted data: Control functionalities, i.e., all communication done between ZTA components, reside on the "control plane", while all other data transfer happens on the "data plane".

The architectural aspects of ZTA have been studied extensively in the past. NIST, in [1], defined four major ZTA types together with requirements for ZTA components, use-cases for ZTA and possible threats. We exclude the "Resource Portal" and "Device Application Sandboxing" models, as they do not incorporate a client side agent. Instead, we focus on the classic "Enclave" and "Device Agent/Gateway" deployment models. Figure 1 shows a generic ZTA based on the latter model, with ZTA specific components depicted in dark blue. It contains the four main components, i.e., a Policy Information Point (PIP), Policy Enforcement Point (PEP), Policy Decision Point (PDP), an agent, a subject accessing a service via an endpoint and further data sources. In addition to the components themselves, we identified five communication flows between components which are specific to ZTA. They are numbered and discussed in the following sections at the affected components. The remaining, unnumbered flows are not specific to ZTA and can be realized with existing protocols and APIs.

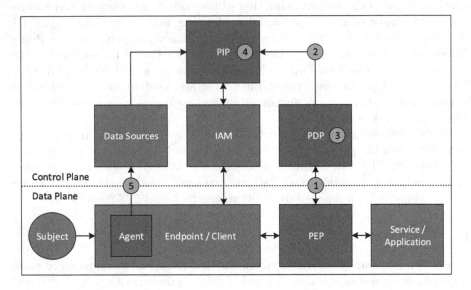

Fig. 1. Generic Zero Trust Architecture with all communication flows

4.2   Federation

In a federated system, multiple distinct entities or organizations collaborate to form a new, larger entity. Trust between participating organizations needs to be established beforehand. In the context of ZTA, federation can be achieved by connecting multiple independent ZTAs to allow seamless data exchange.

User, asset and policy-related information must be exchanged for federated operation. When multiple ZTA's collaborate in this way, that set of information is extended – while all other aspects of ZTA stay essentially the same. PDPs and possibly PIPs need to support this mode of operation, e.g., the sharing of data between parties. From an architectural perspective, ZTA is therefore well suited for federated operation.

4.3   Policy Enforcement Point

The PEP is responsible for enforcing access decisions made by the PDP. From a network perspective, the PEP must be located anywhere between the source and the target of every connection. Moving the PEP closer to the target can be advantageous, as the scope of the trust zone is reduced or even eliminated in case of direct integration in the target application.

We evaluate PEPs based on the following three requirements: (1) The architecture, i.e., integrated into the target application, as a proxy component between client and service, and client-side. (2) The protocol used for interacting with the PDP is also a requirement, see Flow 1 in Fig. 1. (3) Push-based demotion and termination of sessions are advanced capabilities a PEP can support and therefore another requirement.

4.4   Policy Decision Point

The PDP is the central component of every ZTA and is responsible for validating and deciding every single access request. Authentication and authorization need to be considered separately here. The PDP needs to authenticate every request, i.e., validating that the requesting party is actually the one it claims to be. This process is usually delegated to Identity and Access Management (IAM) solutions. Authorization is ensured by evaluating the policy. The decision can then be based on authenticated user and, ideally, device information. Usually, services or applications make some authorization decisions themselves. With ZTA, it is possible to move these authorization decisions to the PDP. This approach allows centralized and fine-grained access control schemes to be realized at the PDP.

We evaluate PDPs based on the following five requirements: (1) Supported policy languages, (2) options for ingesting policies, and (3) mechanisms for policy storage are crucial. (4) As a PDP needs to communicate with PEPs, PIPs, and possibly other PDPs, the respective protocols for these purposes are also evaluated, see Flows 1–3 in Fig. 1. (5) The last requirement is the support for federated operation.

4.5   Policy Information Point

The PIP is another central component in every ZTA. It collects all data needed for making policy decisions from various data sources. This information is then offered to the PDP in a standardized manner. Therefore, the distinction between PDP and PIP is only functional, allowing the PIP to be possibly integrated into the PDP.

We evaluate PIPs based on the following four requirements: (1) Data sources, i.e., all supported means of acquiring data from external systems and databases.

(2) Since PIPs need to interact with them, supported Identity Providers (IdPs) are another central requirement. (3) The query protocol, i.e., the protocol the PIP offers to PDPs is also relevant, see Flow 2 in Fig. 1. (4) PIP-to-PIP communication support, e.g., in distributed or federated environments is the last requirement, see Flow 4 in Fig. 1.

4.6   Agent

The agent is installed at the endpoint or integrated into the operating system. It collects relevant information about the host system, such as the operating system version and device trust related data. This information is collected centrally and can be used by the PDP during policy evaluation. It is important to note that some solutions use the agent to establish connections or tunnels to legacy applications or services. This functionality can also be realized via proxies or standalone software and is not necessarily part of the agent nor a requirement. We argue that it is impossible to realize all fundamental ZTA goals in an agent-less ZTA, a view also expressed by NIST, see [1]. Without this component the PIP gains no knowledge about the state of the requesting device, making device trust based decisions infeasible.

We evaluate three requirements for agents: Capabilities for (1) hardware and (2) software collection, i.e., the types of data it can collect about the host system. (3) The protocol used to transmit this data to PIPs or PDPs is the last requirement, see Flow 5 in Fig. 1.

4.7   Control Plane

Communication within a ZTA can be logically separated into the "data plane" and the "control plane". The actual communication is done on the data plane, while management and control functionalities reside on the control plane. ZTA components must therefore support protocols for communication on the control plane. All protocols need to guarantee integrity, confidentiality, and reliability. Non-functional requirements such as performance are out of scope for this work.

To the best of our knowledge, the current protocols for these communication flows are either proprietary or custom-developed and tightly integrated into the respective software solution without ongoing standardization efforts. Due to this tight coupling, the protocols are evaluated as part of the component in question.

5   Implementations and Protocols

Our systematic analysis of ZTA software solutions and protocols suited for usage on the control plane resulted in several implementations and emerging standards that we discuss in the following sections. For each solution, we analyze and discuss use-cases, implemented components, supported features, interfaces, and finally, security properties of employed protocols.

The **Envoy Proxy** project is a proxy component for micro-services, i.e., cloud-native applications [21]. The software is supposed to be installed on the application server and secures communication between applications by tunneling via mTLS connections. Envoy can tunnel any TCP/UDP traffic. In addition, a variety of application protocols are supported, for example, HTTP, Redis, and Postgres. While the primary use-case is securing service-to-service communication, Envoy can also be used as an edge proxy that accepts requests from clients and forwards those to services. The authorization decision for incoming requests can be delegated to external components via custom filters. Envoy can therefore be combined with other zero trust solutions to work as a PEP and can be deployed on Linux, macOS, Windows, and Docker containers. Envoy is licensed under the Apache License 2.0.

OpenZiti offers a complete zero trust solution comprising all necessary components: A PEP (EdgeRouter), a PDP (Controller), a PIP (integrated into the controller), a client software (Client) and a custom developed control plane protocol [22]. Endpoints must use the Client to access applications secured with OpenZiti. EdgeRouters form a mesh and are able to tunnel TCP or UDP-based protocols over untrusted networks. The final EdgeRouter in front of the target application (from a networking point of view) terminates the secure tunnel. In addition to the tunneling mechanism, OpenZiti offers SDKs for C, C#, Swift, and REST. Applications built on top of the SDK can directly interface with the OpenZiti network without needing to terminate the tunnel. The access policy is configured at the Controller. It offers "posture checks" that can be used for authentication and device assurance: (1) operating system type and version, (2) network adapter MAC address, (3) external MFA with a configurable timeout, (4) running applications defined by the path of the executable, and (5) windows domain membership. All connections inside an OpenZiti network are secured using mTLS with X.509 certificates. OpenZiti can be deployed on Linux, Windows, macOS and Docker containers. The client component additionally supports Android and iOS. All components except the client applications for proprietary operating systems are developed as open-source software and licensed under the Apache License 2.0.

Tailscale is a modern zero trust capable VPN solution that can be used to build secure tunnels to services and applications over untrusted networks [23]. Tailscale consists of the client software installed at every host that should be part of the network and the central server software. The clients form a mesh network that securely tunnels traffic. The central coordination server is hosted by Tailscale Inc. and coordinates the distribution of authentication keys and access policies. A third-party open-source implementation of the coordination server called "Headscale" is also available [24]. More complex setups are also possible. For instance, Tailscale is able to construct complete VPN tunnels that encompass all traffic originating from the client, so-called exit nodes. Tailscale is based on Wireguard, a modern VPN protocol and software, see [25]. The client software is available for Linux, BSD, Windows, macOS, iOS, and Android. The coordination server software is available for Linux, BSD, Windows, and macOS.

Tailscale is licensed under the BSD 3-Clause license while the client software user interface code for Windows, macOS, and iOS is proprietary.

Pomerium is an access proxy solution comprised of four components: (1) the proxy service, responsible for tunneling all traffic, (2) the authentication service that connects the IdP to the system and manages session cookies, (3) the authorization service that validates every request against a dynamic access policy, and (4) the "Data Broker Service" that stores session information, identity data, and tokens [26]. The authorization and authentication components check the context of the request, the requesting user's identity and the device identity. Pomerium supports all HTTP-based traffic and allows tunneling arbitrary TCP-based traffic via HTTP. Applications managed by Pomerium receive the user's identity via a signed JWT. The JWT's signature can then be verified by the application. Communication between the components is using gRPC secured with X.509 certificates. Communication between Pomerium and services or applications is secured via mTLS with X.509 certificates. Pomerium is licensed under the Apache License 2.0 and can be deployed on Linux, macOS, and Docker containers.

Boundary is an identity-based access solution consisting of the client software, "Controller" and "Worker" nodes [27]. Both types of nodes can operate redundantly for scaling and failover functionality. Client access requests are authenticated via a controller node, while worker nodes act as proxies for the actual application data. The communication between the service or application and worker nodes is unencrypted, as the secure tunnel is terminated at the worker node. Communication between nodes is based on mTLS secured with X.509 certificates. Boundary can be deployed on Linux, Windows, and macOS and is licensed under the Mozilla Public License 2.0.

Ockam is a library for secure end-to-end communication [28]. It serves as a PEP and can be used to build zero trust capable Rust applications. Key establishment, rotation, and revocation as well as attribute-based access control mechanisms are supported. Ockam supports TCP, UDP, Websockets and Bluetooth as transport protocols. Ockam applications establish secure channels via the Noise Protocol Framework or the X3DH protocol [29]. Ockam is licensed under the Apache License 2.0.

Oathkeeper is a zero trust capable HTTP reverse proxy with an integrated decision API [30]. It can be used as a combined PEP and PDP while also supporting PDP functionality in standalone mode. The reverse proxy ensures that requests satisfying the access rule pipeline are forwarded to the upstream server. The access rule pipeline consists of four components: (1) Authentication handlers inspect and validate HTTP traffic using sessions (cookie-based), tokens, OAuth 2.0 or JWTs. (2) Authorization handlers check access permissions based on Ory Keto or arbitrary remote HTTP and JSON endpoints. (3) Mutation handlers can be used to transform and augment authentication information into formats the authentication backend understands. This includes creating signed JWTs, arbitrary HTTP header data and cookies. Authentication information can also be enriched by querying external APIs. (4) Error handlers that define behavior in case authentication or authorization fails. Options include JSON responses, HTTP redirects, and HTTP 401 "WWW-Authenticate" responses. The access control API can also be connected to Ambassador, the Envoy proxy, AWS or Nginx.

Keto, see [31], is an open-source implementation of Zanzibar [32], an authorization system developed by Google. It can be used as a PDP. Keto stores the policy as relation tuples between subjects and objects. Relation tuples form a graph from which permissions can be deducted. Relations can be queried via HTTP and gRPC API endpoints, allowing read operations to check permissions, query relations, and list objects. Write operations can be used to modify, insert and delete objects and relations. Access to Oathkeeper and Keto APIs is secured using HTTPS. Both solutions can be deployed on Linux, macOS, Windows, and FreeBSD and are licensed under the Apache License 2.0.

The **Open Policy Agent** (OPA) is a general-purpose, open-source policy engine [33] that can be used as a PDP. Together with the declarative "REGO" policy language, it allows the creation and execution of fine-grained policies that can be used to build ZTAs. OPA offers an HTTP REST API for evaluating policies that returns JSON data. In addition, OPA can be integrated into Go applications via an SDK. Policies can also be compiled into WebAssembly instructions and can be embedded into any WebAssembly runtime. OPA can be deployed on Linux, macOS, Windows, and via Docker containers. OPA is licensed under the Apache License 2.0.

The **Secure Production Identity Framework For Everyone** (SPIFFE) and the accompanying **SPIFFE Runtime Environment** (SPIRE) are a standard and reference implementation for identification, attestation and certificate distribution in dynamic software systems [34]. The SPIFFE standard defines SPIFFE IDs that can be used as identities for services. These IDs can be encoded into SPIFFE Verifiable Identity Documents (SVIDs), cryptographically verifiable documents, i.e., certificates. SPIFFE also defines an API for issuing and retrieving SVIDs called the "Workload API". SPIRE implements SPIFFE and offers additional features. SPIRE consists of agents installed alongside the application or service and a server component. SPIRE can perform node and workload attestation and registration as well as rotate keys and certificates. It also provides services with access to secret stores and databases and enables federation of SPIFFE systems across trust boundaries.

SPIREs API endpoint is offered by the agent and used by a workload for creating and validating SVIDs. Communication uses gRPC over a Unix Domain or TCP socket. As SPIFFE is often used to establish a root of trust, TLS must not be required by implementations. The communication between the agent and the central server component is secured using mTLS with a pre-shared "bootstrap bundle".

In the context of ZTA, SPIFFE/SPIRE can be used as part of the control plane. It can be used to quickly and securely retrieve, rotate, and manage service identities and corresponding keys in cloud environments. In this context, containers for services are usually rapidly deployed and decommissioned. This is the primary use-case for SPIFFE/SPIRE, as requirements differ from traditional mechanisms for identity management of client devices and servers. It is licensed under the Apache License 2.0.

The **Shared Signals Framework** (SSF) is currently being developed by the "Shared Signals Working Group" at the OpenID foundation and aims to standardize a security event sharing protocol [35]. The OpenID foundation plans to develop and provide a reference implementation for SSF to facilitate interoperability testing. Two profiles have been developed for SSF: The Continuous Access Evaluation Protocol (CAEP) and the Risk Incident Sharing and Coordination (RISC). **CAEP** was proposed by Google [16] and later merged into SSF. Without Continuous Access Evaluation, access decisions are only made once before establishing connections. To fully realize the benefits of ZTA, access decisions must be evaluated continuously. To that end, CAEP standardizes events for communicating access property changes between zero trust components. For example, the "Device Compliance Change"-Event can signal that a device no longer fulfills an organization's security policy, possibly including a reason. Such an event could be generated by an agent, received by a SIEM, and later taken into account during policy evaluation at the PDP. **RISC** is the second profile in development. It focuses on transmitting events and information concerning user account security. While not an integral part of ZTA, it can be used for collaboration in federated scenarios. Events have been specified to signal changes related to accounts, credentials and recovery information. For example, RISC can be used to prevent attackers from using compromised credentials several times at different providers. As SSF is built upon Security Event Tokens (SET), see [36], TLS 1.2 or higher must be supported for the transport of events. In addition, SETs must be encrypted in case they contain personally identifiable information (PII) and must ensure integrity, for example by using JWS [37].

Other solutions and commercial offers were found during our search but not included in the evaluation: "beyond" [38] and "helios" [39] are zero trust HTTP access proxy solutions. Due to missing documentation and stalled development, these solutions were not considered. "TRASA", see [40], is an identity and context aware access proxy. It can be used to secure remote access to internal services. TRASA supports RDP, SSH, HTTP, and MySQL. Since the development of TRASA has stopped in December 2021, it was excluded from the evaluation. "Pritunl Zero", see [41], is a zero trust proxy solution for SSH and HTTP connections. While under active development, this software is licensed under a custom license that only allows non-commercial use and forbids the distribution of derivative works. It primarily serves the use-case of centralizing the management of SSH keys through a custom-built certificate authority and a custom SSH client. The offered authentication options are limited, and the solution supports neither device authentication nor assurance. The developers offer a commercial and proprietary solution called "Pritunl" with extended features and support. For these reasons, Pritunl Zero was excluded from the evaluation.

In addition to the open-source solutions described and discussed in the previous sections, we discovered multiple commercial zero trust offers. Google

(BeyondCorp Enterprise) and Microsoft (integrated into Microsoft 365) are the primary vendors in this context. Both are cloud-based and offer advanced solutions for zero trust security. These solutions are proprietary, hosted and operated by third parties. Therefore, their capabilities cannot be evaluated like it is the case with open-source software and are of lesser academic interest. Finally, NIST is working on a practice guide titled "Implementing a Zero Trust Architecture" in cooperation with industry partners, see [42]. The goal is to introduce a reference ZTA built on commercially available technology.

6   Evaluation

A comparison of the nine ZTA software systems analyzed in this work is shown in Fig. 2. It is based on the generic ZTA constructed in Sect. 4. The implemented components are highlighted for every software system investigated in this work. The following evaluation is done on a per-component basis for the requirements defined in Sect. 4 and listed in Appendix A. General findings and directions for future work are discussed in Sect. 7.

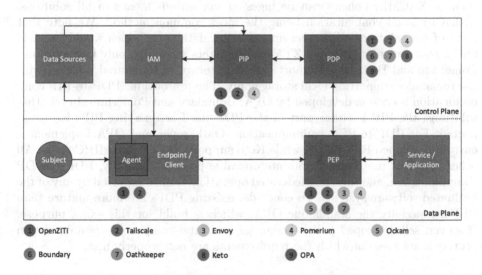

Fig. 2. Component-wise comparison of ZTA software

6.1   Policy Enforcement Point

Existing solutions support all three architectures for PEPs. Support for proxy mode is implemented in all solutions except Ockam, a library that can be used to build zero trust capable applications. Integrated and client-side PEP operation is only provided by two solutions. This result can be attributed to the fact that many solutions build tunnels for the transmission of data plane and inter-component communication. The protocol to connect PEPs to PDPs is self-developed in nearly all solutions. This prevents interoperability, i.e., mixing components from different vendors. The only exception is the Envoy proxy project, a standalone PEP implementation with vast filtering and processing capabilities for requests. Envoy can query PDPs via gRPC and standalone HTTP calls, making it the most capable solution in this regard. Finally, push-based demotion or termination of connections is not documented for any solution. We assume that this advanced capability was not prioritized in the past. We conclude that existing PEPs partially meet the requirements we defined, although interoperability and non-tunnel based operation is lacking.

6.2   Policy Decision Point

The primary requirement for PDPs is the language in which policies are expressed. Only the Open Policy Agent uses a well-defined language for this purpose (REGO), while all other solutions implement custom languages. It is unclear why most examined solutions do not implement established policy languages such as XACML. Policies can be ingested via various means in all solutions, with CLI-based configuration being the most common method. We note that none of the solutions allow querying external databases, which would be ideal for large-scale and distributed ZTA environments. Similarly, only two solutions, Pomerium and Boundary, support external storage of configured policies, with the remainder supporting local storage only. The protocol for PDP-to-PIP communication is custom developed by OPA, Boundary and Pomerium. In all other solutions, the PIP is either part of the PDP itself or querying PIPs is unsupported. For PDP-to-PEP communication, Oathkeeper and OPA implement a custom-developed REST API, while Keto supports HTTP and gRPC APIs. All other solutions use custom, non-interoperable protocols. Finally, PDP-to-PDP communication, and therefore federated operation, is not supported by any of the evaluated software systems. To conclude, existing PDPs are more mature than PEPs, especially the standalone OPA, which is build for this exact purpose. However, self-developed policy languages and inter-component communication protocols are areas in which the requirements are not properly met.

6.3   Policy Information Point

We were unable to locate standalone implementations for PIPs. Instead, the examined solutions implement the PIP as part of the PDP. This can serve as an explanation to why no solutions supports external data sources apart from Tailscale, Headscale, and Pomerium, which are able to use authentication data from IdPs. They support a vast number of IdPs, among them G Suite, Azure, and generic OIDC or SAML providers. In addition, all solutions use an internal RBAC model for configuring an access policy. None of the solutions support a protocol for connecting to PDPs, because they merge the PDP and PIP components allowing data exchange to happen internally. For the same reason, federated operation is also unsupported in all solutions. This allows us to conclude that the requirements for PIPs are not fulfilled in any solution.

6.4 Agent

Data collection about the host system's software and hardware is the main functionality for agents. None of the solutions fulfill these requirements in a strict sense, i.e., collecting and transmitting this data to an external component such as a PIP. Instead, agent implementations in Boundary, OpenZiti, and Tailscale establish tunnel connections to virtual overlay networks which terminate at PEPs. OpenZiti is the only solution with rudimentary device assurance checks during connection establishment, e.g., validating operating system type and version. This also results in custom developed protocols for Agent-to-PIP communication in solutions that implement this feature. Microsoft's zero trust solution encompasses an agent component called "Intune" that is integrated in the Windows operating system. It is a notable exception with regard to the data collection requirement, as vast information about the state of both software and hardware of the host can be collected and evaluated centrally. As Intune is a proprietary, commercial solution, it can not be used in combination with third party ZTA components. We conclude that in a strict sense, no open-source implementations of agents exist, and therefore, none of the requirements are fulfilled.

7 Challenges and Future Work

In Sect. 5 we discussed the various aspects of current software solutions and protocols for ZTA. The requirements for the evaluation were explained in Sect. 4 and matched to the solutions in Sect. 6. Based on our detailed analysis, we can identify the different challenges we have observed and formulate directions for future work in this field. In this section, we discuss them in detail.

Interoperability: Our current networks are based on open standards that allow communication between individual solutions. This interoperability was crucial for the success of many network technologies. In the field of ZTA, we currently see many proprietary solutions that do not allow this interoperability between implementations. Complete solutions that implement all four primary ZTA components such as OpenZiti, Tailscale, and Pomerium offer few or no means to interoperate with other solutions. Partial solutions such as Envoy or OPA implement specific components and are better positioned in this regard by offering custom API endpoints. Nevertheless, interoperability is a significant challenge with existing solutions and a promising direction for future work. Standards for inter-component communication need to be standardized and implemented.

Control Plane: We assess that no open standards or protocols exist specifically for communication between ZTA components on the control plane. This applies to both communication between individual components deployed as part of the same ZTA, for example, between a PEP and the PDP, and inter-organizational communication in the federated scenario, i.e., between PDPs. Promising standardization efforts such as SSF are ongoing. SSF aims to address inter-organizational and inter-service security event sharing. This is an essential step towards improving the overall security posture of IT systems and not limited to ZTA. These protocols and standards need to be developed in the future. Research in this direction can foster the adoption of ZTA by allowing cross-vendor component compatibility.

Federation: As discussed in Sect. 4.2, ZTA is an ideal candidate for federated operation. We notice that none of the solutions we assessed support federation, which is not surprising given the lack of suitable control plane protocols. Academic work concerning this topic is also sparse. As the adoption of ZTA progresses, the importance of federated operation will rise accordingly. We therefore see the possibility for future work in federated architecture, policy evaluation in federated systems, and privacy-related aspects.

Device Trust and Assurance: Existing solutions are well-positioned with regard to user authentication. However, device identification, trust, and assurance are severely lacking, as none of the solutions support an agent with the necessary capabilities. Including information about client devices in the decision process of the PDP is one of the main tenets of ZTA. This provides a variety of directions for future work. For example, research can focus on how to acquire this data in specific scenarios such as enterprise IT or cloud setups. Extending this idea to diverse environments with limited hardware capabilities, such as the IoT or OT space, is another possible direction. Protocol-based support can also be considered, for example by integrating device identification and assurance checks in standard authorization protocols such as OpenID Connect [43].

Policy Information Point: While PIPs as a concept are well-defined and referenced by the literature, not a single standalone open-source PIP exists today. Complete ZTA solutions offer PIPs with rudimentary functionality, falling short of what is required by a wide margin. Specifically, interfacing with external data sources other than IdPs, e.g., SIEMs or CTI systems, is not supported by any solution. This hinders the adoption of ZTA as a PIP offers functionality central to the idea of ZTA. Future work should examine why this is the case and focus on developing solutions.

Agent: Challenges related to the agent component are similar to PIPs. First, standalone agent components do not exist. Instead, they are implemented as part of a larger solution as it is the case with Tailscale or OpenZiti, offering only basic agent functionality. Microsoft offers an agent that is integrated into the Windows operating system as part of their ZTA solution. We believe this is the right approach to the problem since agents running as standalone components in user space suffer from several drawbacks. For adequate visibility and control functionality, they must run with elevated privileges and interface with the operating system to acquire the necessary data. It would therefore be ideal to integrate this functionality directly into the operating system. This task relies on the vendor for proprietary systems such as Windows, macOS and iOS. Future work can focus on how to securely acquire and collect this information in open systems such as Linux and BSD derivatives, for example, by integrating it directly into the kernel.

Scaling: One crucial factor in deploying ZTA within an organization's IT network is the topic of scaling. IT infrastructures must be globally reachable with low latency and capable of handling a large number of user sessions simultaneously. Of the examined solutions, six support standalone operation only. Therefore, we propose to look into this functionality in current and future implementations.

8 Conclusions

Zero Trust Architecture is a network security paradigm in which trust, and thus access, is explicitly granted based on user and device authentication. As an organization's perimeter can no longer be clearly defined and protected, the location of the user and the device does not impact authorization decisions within ZTA. While is an emerging field of research with accelerating, commercial deployment and usage, the state of open-source solutions is not well understood. In this work, we surveyed available open-source ZTA solutions and discussed requirements for ZTA software components. Based on these, we evaluated how mature the open-source software components are.

Our findings show that some solutions in the zero trust space aim to implement an "all-in-one" solution, i.e., all components necessary to deploy a ZTA. In contrast to these, there are standalone components that are capable of interfacing with third-party software, allowing modular ZTAs to be built. With regard to protocols, the "Shared Signals Framework" standardization effort stands out as a major development for an open control plane protocol. We identified seven key challenges that offer the potential for future work: Component interoperability, control plane protocols, and federated ZTA are still in the early stages. Concerning components, we assess that standalone Policy Information Points and agents are not available as of today. Device trust and assurance functionalities are essential to ZTA but need to be improved going forward, as existing solutions barely support them. Lastly, open questions regarding scaling in ZTA also need to be addressed.

To conclude, we assess that in general, open-source solutions for ZTA are not yet mature. Protocols for ZTA are even less developed than the software side.

Data Availability. The full results of the evaluation, i.e., a table listing the components together with our assessment of fulfilled requirements, is available online [44].

Competing interests. All authors declare that they have no conflicts of interest.

A   List of Requirements

In the following, we list the requirements used in the evaluation together with a short explanation, examples and the desired state.

A.1   Policy Enforcement Point

- **Architecture:** *Integrated operation,* i.e., directly integrated into the target application or service, for example via libraries. *Proxy mode,* i.e., as a seperate component in front of the server. *Client side,* i.e., the PEP is deployed on the client device. All architectures are equally desirable.
- **Supported protocols for PEP-PDP communication:** A list of protocols the PEP supports for interacting with PDPs. Existing, well-defined protocols with wide usage are desirable.
- **Push-based demotion and termination:** This advanced feature allows PDPs to demote or terminate established sessions by instructing the PEP to do so. Support for this is desirable.

A.2   Policy Decision Point

- **Supported policy languages:** A list of languages the PDP supports for encoding policies. Existing, well-defined languages with wide usage are desirable.
- **Options for ingesting policies:** A list of options allowing the ingestion of policy information. Examples include user interfaces, REST APIs and the file system. CLI- or API-based ingestion is preferred.
- **Policy storage mechanisms:** A list of supported ways to store policies. The local filesystem is an example. Here it is desirable to have the option of using a database.
- **Supported protocols for PEP, PIP, PDP communication:** A list of protocols the PDP supports for interacting with other components. Existing, well-defined protocols with wide usage are desirable.
- **Federated operation:** The capability and maturity of operating the PDP in a federated environment. It is desirable to have this feature.

A.3   Policy Information Point

- **Data sources:** A list of supported data sources the PIP can query. Examples include device registries, generic databases and CTI feeds. Support for many sources is desirable.
- **Identity providers:** A list of IdPs the PIP can interface with, for example generic support for SAML. Support for well known IdPs and protocols is desirable, especially OIDC.
- **Query protocol:** The protocol, or the protocols, the PIP supports for querying data. This protocol can then be used by the PDP for policy decisions. Existing, well-defined protocols with wide usage are desirable.
- **Supported protocols for PIP-PIP communication:** In federated environments, PIPs might need to interface with other PIPs to exchange data. Existing, well-defined protocols with wide usage are desirable.

A.4   Agent

- **Hardware-based collection capabilities:** A list of information items the agent is able to collect about the hardware of the client. Examples include the secure boot state, firmware version information or CPU vulnerabilities. It may be desirable to collect as much information as possible.
- **Software-based collection capabilities:** A list of information items the agent is able to collect about the software of the client. Examples include the operating system type and version, currently running software or antivirus software state. It may be desirable to collect as much information as possible.
- **Supported protocols for Agent-PIP communication:** A list of protocols the agent supports for interacting with PIPs. Existing, well-defined protocols with wide usage are desirable.

References

1. National Institute of Standards and Technology: [NIST SP 800–207] Zero Trust Architecture. NIST Special Publication - 800 series (2020)
2. Jericho Forum ™ Commandments (2007). https://collaboration.opengroup.org/jericho/commandments_v1.2.pdf
3. Ward, R., Beyer, B.: BeyondCorp : a new approach to enterprise security. Login Mag. USENIX & SAGE **39**(6), 6–11 (2014)
4. Zimmer, B.: LISA: a practical zero trust architecture. In: Enigma 2018 (Enigma 2018). USENIX Association, Santa Clara, CA (2018)
5. Microsoft Corporation: Zero Trust Model - Modern Security Architecture|Microsoft Security (2022). https://www.microsoft.com/en-us/security/business/zero-trust. Visited 13 Apr 2023
6. Microsoft and Hypothesis Group: Zero Trust Adoption Report (2021)
7. Kindervag, J.: No More Chewy Centers: Introducing the Zero Trust Model of Information Security (2010)
8. The Open Group: Zero Trust Core Principles (2021). https://publications.opengroup.org/w210

9. Rose, S.: Planning for a Zero Trust Architecture: A Planning Guide for Federal Administrators (2022)
10. Køien, G.M.: Zero-trust principles for legacy components: 12 rules for legacy devices: an antidote to chaos. Wireless Pers. Commun. **121**, 1169–1186 (2021). https://doi.org/10.1007/s11277-021-09055-1
11. He, Y., Huang, D., Chen, L., Ni, Y., Ma, X.: A survey on zero trust architecture: challenges and future trends. Wirel. Commun. Mobile Computing **2022**, 1–13 (2022)
12. Buck, C., Olenberger, C., Schweizer, A., Völter, F., Eymann, T.: Never trust, always verify: a multivocal literature review on current knowledge and research gaps of zero-trust. Comput. Secur. **110**, 102436 (2021). https://doi.org/10.1016/j.cose.2021.102436
13. Syed, N.F., Shah, S.W., Shaghaghi, A., Anwar, A., Baig, Z., Doss, R.: Zero Trust Architecture (ZTA): a comprehensive survey. IEEE Access **10**, 57143–57179 (2022). https://doi.org/10.1109/ACCESS.2022.3174679
14. Olson, K., Keller, E.: Federating trust: network orchestration for cross-boundary zero trust. In: Proceedings of the 2021 SIGCOMM 2021 Poster and Demo Sessions, Part of SIGCOMM 2021 (2021). https://doi.org/10.1145/3472716.3472865
15. Hatakeyama, K., Kotani, D., Okabe, Y.: Zero trust federation: sharing context under user control towards zero trust in identity federation. In: 2021 IEEE International Conference on Pervasive Computing and Communications Workshops and other Affiliated Events, PerCom Workshops 2021 (2021). https://doi.org/10.1109/PerComWorkshops51409.2021.9431116
16. Tulshibagwale, A.: Re-thinking federated identity with the Continuous Access Evaluation Protocol|Google Cloud Blog (2019). https://cloud.google.com/blog/products/identity-security/re-thinking-federated-identity-with-thecontinuous-access-evaluation-protocol. Visited 13 Apr 2023
17. Maler, E., Machulak, M., Richer, J., Hardjono, T.: User-Managed Access (UMA) 2.0 Grant for OAuth 2.0 authorization. Technical report (2019)
18. Hardt, D.: The OAuth 2.0 Authorization Framework. RFC 6749, RFC Editor (2012)
19. Wohlin, C.: Guidelines for snowballing in systematic literature studies and a replication in software engineering. In: Proceedings of the 18th International Conference on Evaluation and Assessment in Software Engineering, pp. 1–10 (2014)
20. Anderson, A., et al.: eXtensible Access Control Markup Language (XACML) Version 2.0. Oasis (2004)
21. Envoy Project Authors: envoyproxy/envoy: Cloud-native high-performance edge/middle/service proxy (2016). https://github.com/envoyproxy/envoy. Visited 13 Apr 2023
22. Styra Inc.: OpenZiti: programmable network overlay and associated edge components for application-embedded, zero-trust networking (2019). https://github.com/openziti/. Visited 13 Apr 2023
23. Fond, J.: Juanfont/Headscale: An Open Source, Self-Hosted Implementation of the TAILSCALE Control Server (2020). https://github.com/juanfont/headscale. Visited 13 Apr 2023
24. Tailscale Inc.: Tailscale is a WireGuard-based app that makes secure, private networks easy for teams of any scale (2020). https://github.com/tailscale. Visited 13 Apr 2023
25. Donenfeld, J.A.: Wireguard: next generation kernel network tunnel. In: NDSS 2017, pp. 1–12. The Internet Society (2017)

26. Pomerium Inc.: pomerium/pomerium: Pomerium is an identity-aware access proxy (2019). https://github.com/pomerium/pomerium. Visited 13 Apr 2023
27. HashiCorp Inc: hashicorp/boundary: Boundary enables identity-based access management for dynamic infrastructure (2020). https://github.com/hashicorp/boundary. Visited 13 Apr 2023
28. Ockam Inc.: build-trust/ockam: Orchestrate end-to-end encryption, mutual authentication, key management, credential management & authorization policy enforcement - at scale (2018). https://github.com/build-trust/ockam. Visited 13 Apr 2023
29. Marlinspike, M., Perrin, T.: The X3DH key agreement protocol. Open Whisper Syst. **283**, 10 (2016)
30. Ory Corp: ory/oathkeeper: a cloud native Identity & Access Proxy/API (IAP) and Access Control Decision API that authenticates, authorizes, and mutates incoming HTTP(s) requests (2017). https://github.com/ory/oathkeeper. Visited 13 Apr 2023
31. Ory Corp: ory/keto: Open Source (Go) implementation of "Zanzibar: Google's Consistent, Global Authorization System" (2018). https://github.com/ory/keto. Visited 13 Apr 2023
32. Pang, R., et al.: Zanzibar: Google's consistent, global authorization system. In: 2019 USENIX Annual Technical Conference (USENIX ATC 2019), Renton, WA (2019)
33. Styra Inc.: open-policy-agent/opa: An open source, general-purpose policy engine (2015). https://github.com/open-policy-agent/opa. Visited 13 Apr 2023
34. The SPIFFE authors: SPIFFE: Secure Production Identity Framework for Everyone (2017). https://spiffe.io. Visited 13 Apr 2023
35. OpenID Foundation: Shared Signals - A Secure Webhooks Framework|OpenID (2017). https://openid.net/wg/sharedsignals/. Visited 13 Apr 2023
36. Hunt, P., Jones, M., Denniss, W., Ansari, M.: Security Event Token (SET). RFC 8417, RFC Editor (2018)
37. Jones, M., Bradley, J., Sakimura, N.: JSON Web Signature (JWS). RFC 7515, RFC Editor (2015)
38. cogolabs contributers: cogolabs/beyond: BeyondCorp-inspired HTTPS/SSO Access Proxy. Secure internal services outside your VPN/perimeter network during a zero-trust transition (2017). https://github.com/cogolabs/beyond. Visited 13 Apr 2023
39. Yakimov, C.: cyakimov/helios: Identity-Aware Proxy (2019). https://github.com/cyakimov/helios. Visited 13 Apr 2023
40. Seknox Pte. Ltd.: seknox/trasa: Zero Trust Service Access (2020). https://github.com/seknox/trasa. Visited 13 Apr 2023
41. Pritunl Inc: pritunl/pritunl-zero: Zero trust system (2017). https://github.com/pritunl/pritunl-zero. Visited 13 Apr 2023
42. Kerman, A., Souppaya, M., Grayeli, P., Symington, S.: Implementing a zero trust architecture (Preliminary Draft), Technical report, National Institute of Standards and Technology (2022)
43. Sakimura, N., Bradley, J., Jones, M., De Medeiros, B., Mortimore, C.: OpenID Connect Core 1.0. The OpenID Foundation (2014)
44. Hilbig, T.: hm-seclab/paper-th-zta-components-materials: Supporting materials for STM 2023 (2023). https://github.com/hm-seclab/paper-th-zta-componentsmaterials. Visited 19 Aug 2023

Application Scenarios

Secure Stitch: Unveiling the Fabric of Security Patterns for the Internet of Things

Emiliia Geloczi[(✉)] [ID], Felix Klement[(✉)] [ID], Eva Gründinger,
and Stefan Katzenbeisser

University of Passau, Innstraße 43, 94032 Passau, Germany
{emiliia.geloczi,felix.klement,stefan.katzenbeisser}@uni-passau.de,
gruendinger@fim.uni-passau.de

Abstract. The design of the Internet of Things (IoT) system is a complex process, not only in terms of the balance between resource consumption and extensive functionality but also in the context of security. As various technical devices are now widespread and have access to all kinds of critical information, they become one of the main targets for attackers. Consequently, it is vital to consider the IT security aspect during the development of any system. A practical way to do it is to use security patterns. There are many different patterns that can address particular problems, but not all of them are suitable due to the wide range of requirements in such systems. In this paper, we present a systematic collection and categorisation of IoT-applicable security patterns and analyse gaps in recent research works related to security. We provide a catalogue of 61 patterns organised in a top-down approach that follows the World Forum's IoT Architecture Reference Model, this collection is able to play an important role in the future development of secure IoT solutions.

Keywords: IoT · IoT Security · Design Patterns

1 Introduction

The proliferation of the Internet of Things (IoT) has ushered in a transformative paradigm wherein countless devices are interconnected, facilitating seamless communication and data exchange. Spanning domains such as smart homes, wearables, industrial systems, and smart cities, the IoT offers unparalleled convenience, efficiency, and connectivity. However, the extensive connectivity inherent in the IoT landscape also introduces significant security challenges [33]. As

This work has been partially funded by the Bavarian Ministry of Science within the framework of the research cluster "ForDaySec: Security in everyday digitalisation", as well as, by the German Federal Ministry of Education and Research, as part of the Project "6G-RIC: The 6G Research and Innovation Cluster" (project number 825026). E. Geloczi and F. Klement—Contributed equally to this work and share first authorship.

the IoT ecosystem expands, so does the attack surface, rendering it susceptible to an array of threats encompassing privacy breaches, data manipulation,
physical harm, and critical infrastructure disruption. Addressing these security
concerns assumes paramount importance in guaranteeing the trustworthiness
and dependability of the IoT.

In the field of software development, the utilization of pre-established design
patterns is a prevalent practice for addressing recurring issues. These patterns
serve the purpose of not only circumventing known problems but also guaranteeing seamless integration and support for systems [35]. Nevertheless, not all
security patterns hold the same level of applicability within the realm of the
IoT, due to the presence of numerous and ever-evolving requirements [20]. Consequently, the adoption of more specialized patterns significantly diminishes the
pool of suitable patterns tailored specifically for the IoT domain.

This paper introduces a comprehensive compilation of systematically organized design patterns that pertain to the mitigation of security challenges in the
realm of the IoT. First, we identify existing design patterns according to the several chosen criteria. Then, the patterns are ranked based on seven architecture
levels, five fundamental security objectives and ten common vulnerabilities. As
a result, to our best knowledge, we provide the most comprehensive catalogue of
design patterns suitable for solving security problems in the IoT, which consists
of 61 elements and is organised according to a top-down approach. After the
analysis of the catalogue, we discover that the included patterns cover all layers
of the IoT architecture to varying degrees, address all considered security goals,
and can also be used to mitigate the most common vulnerabilities.

The rest of the paper is organised as follows. Section 2 includes background
information related to terms used in the catalogue. An overview of related works
is presented in Sect. 3. In Sect. 4, the overall methodology is described including
search and selection procedures. The resulting IoT security pattern catalogue is
shown in Sect. 5. Section 6 contains the evaluation and discussion of the obtained
results. The possible application of the presented catalogue is described in Sect. 7.
Finally, Sect. 8 concludes the paper.

2   Background

To facilitate the reader's initiation into the Security Pattern discourse, we have
synthesized the most important aspects. Within this section, we present a concise explanation of the terminology employed in formulating our comprehensive
design pattern catalog.

2.1   World Forum Architecture Layers

During the creation of the catalogue, we classify the patterns according to the
possible architecture levels at which they can be applied. For this purpose, we use
the generally accepted seven architecture layers according to the World Forum
Reference Model (WFRM) [4] (see Fig. 1).

Fig. 1. Seven IoT architecture levels according to WFRM [4].

2.2 Top Ten Common IoT Vulnerabilities

To determine which vulnerabilities can be addressed by the patterns in the catalogue, we focus on the ten most common issues that have been identified by The Open Web Application Security Project (OWASP) community. The OWASP periodically updates and analyses critical problems regarding building and managing IoT systems. According to the last update, the top ten common IoT vulnerabilities are the following [17].

(T1) Weak Passwords
(T2) Insecure Network Services
(T3) Insecure Ecosystem Interfaces
(T4) Lack of Secure Update Mechanisms
(T5) Use of Insecure or Outdated Components
(T6) Insufficient Privacy Protections
(T7) Insecure Data Transfer and Storage
(T8) Lack of Device Management
(T9) Insecure Default Settings
(T10) Lack of Physical Hardening.

2.3 Security Objectives

An adversary can pursue different goals, such as violating the confidentiality or integrity of information. In compiling our catalogue, we focus on the following five main possible targets of attackers and analyze the design patterns under consideration to determine whether they can ensure the protection of these targets [26]:

- Confidentiality: Data resources or information should be protected against unauthorized disclosure and improper use.
- Integrity: Data resources or information should be protected against unauthorized changes, destruction, or loss.
- Availability: Data resources or information are accessible to authorized users when they are needed.
- Authentication: Before a user can access information or resources, they must prove their identity and permission.
- Authorization: Verification of user permissions to access or use requested resources.

3 Related Work

There are several works describing the relevance of patterns and architectures that focus specifically on IoT. However, besides common design patterns and frameworks, security patterns in this area are still in their early stages of development and documentation. This section gives an overview of existing IoT pattern catalogues.

Reinfurt *et al.* [25] describe specific patterns for designing IoT systems that can be applied to the domain of smart factory systems. These patterns cover different areas and operation modes like device communication and management as well as energy supply types.

Besides design patterns also different architectural styles can be utilized for creating IoT systems. In [15], Muccini *et al.* provide a number of abstract reference architectures. Through the implementation of a systematic mapping study, a comprehensive selection process was undertaken, resulting in the identification of a set of 63 papers from a pool of over 2,300 potential works. The outcomes of this study play a crucial role in the classification of current and forthcoming approaches pertaining to architectural-level styles and patterns in the domain of IoT.

In order to get a better idea of the landscape of patterns and architectures that have accumulated over the years in research, Washizaki *et al.* [33,34] analysed the successes and failures of patterns for IoT systems. The authors acknowledge that the development of IoT-specific patterns and architectures has substantial room for improvement due to limitations in documentation and a scarcity of successfully executed implementations.

Fysarakis *et al.* [9] sketch the SEMIoTICS approach to create a pattern-driven framework that is based on already existing IoT platforms. Aiming to guarantee secure actuation and semi-automatic behaviour, the SEMIoTICS project utilizes patterns to encode dependencies between security, privacy, dependability and interoperability properties of smart objects.

Organized in a hierarchical taxonomy, Papoutsakis *et al.* [18] collect and categorize a set of security and privacy patterns. While giving the reader an overview of security- and privacy-related objectives that are relevant in the IoT domain, the goal of this paper is to match these properties to their corresponding patterns. This usable pattern collection should guide developers to create IoT solutions that are secure and privacy-aware by design.

Over the last three years, Rajmohan *et al.* [21–23] published different papers that review the research work regarding patterns and architectures for IoT security and privacy. Despite rising in the number of publications in this area, there is a shortage of pattern IT security solutions at the Network and Device levels. Whereas the Physical Devices and Controller, Connectivity, and Application layers have the largest number of different security solutions.

Through our comprehensive analysis of the existing body of work, we can draw the conclusion that while there exists a multitude of design patterns applicable to the realm of IoT, there remains a notable dearth of design patterns specifically addressing some security concerns.

4   Methodology

This section introduces the methodology used for the creation of our catalogue. To begin with, we describe the chosen search strategy, and then outline the selection procedure including criteria based on which the founded papers have been filtered.

4.1   Search Strategy

As a base for a search of the existing papers relevant to our topic, we used the Systematic Literature Review (SLR) approach introduced by Kitchenham *et al.* [11]. SLR entails a methodical analysis of publications concerning a specific topic, encompassing the meticulous collection and critical evaluation of multiple research studies or papers. The objective of this study is to offer a comprehensive synthesis of the pertinent literature pertaining to a particular research question, ensuring transparency and reproducibility throughout the process. The following points utilize the review protocol that is used to conduct the literature review in a strategic manner and consist of the research questions that should be answered, selection criteria the found papers need to fulfill and a search strategy on how to browse databases in order to find the most relevant publications.

The strategy to find papers that discuss security patterns is divided into two main parts: automatic and manual search. To conduct our primary search, we employ five widely recognized scientific publication databases: IEEE Xplore, ACM Digital Library, ScienceDirect, Web of Science, and Scopus. Given that Scopus and ACM Digital Library already index SpringerLink, we exclude the former from our search process. Additionally, we omit Researchgate and Google Scholar, as they encompass a considerable number of non-peer-reviewed and non-English papers. Following the predefined set of search engines, we employ an automated approach utilizing specific keywords to identify relevant example studies. Subsequently, we proceed with a manual search to address any potential gaps, ensuring the inclusion of any pertinent scientific papers that may have been overlooked during the automated search process.

Furthermore, we meticulously assess all the obtained papers to determine their adherence to the following criteria:

Inclusion Criteria
IC1 Paper contains (one or more) security pattern that is applicable to an IoT system.
IC2 Paper targets the IoT field, either in a general or specific application domain of IoT.
IC3 Paper discusses security objectives for system design, architecture or infrastructure.
Exclusion Criteria:
EC1 Paper is not written in English language.
EC2 Paper discusses design, privacy or misuse patterns, as well as security architectures not for the IoT domain.

EC3 Paper is not peer-reviewed.

In order for an article to be selected it must meet all inclusion criterion and none of the exclusion criteria:

$$(IC1 \wedge IC2 \wedge IC3) \wedge \overline{(EC1 \vee EC2 \vee EC3)} = 1$$

Lastly, to mitigate the presence of duplicate entries, the final stage of the search process involves the merging of identical papers identified by distinct search engines.

Automatic Search. Following the SLR approach, we identify several keywords in order to create appropriate requests for a successful automatic search. Based on our preceding analysis, we formulate the subsequent search query, which was subsequently employed for our initial database search:

$$(\text{"Internet of Things"} || \text{"IoT"} ||$$
$$\text{"Cyber Physical Systems"} || \text{"Web of Things"})$$
$$\wedge$$
$$(\text{"Security Pattern"} || \text{"Security Design Pattern"})$$

For each search the query string needed to be slightly modified to fit each database's advanced search functionality and guidelines.

Manual Search. By utilizing the snowballing strategy that was introduced by Wholin and Prikladnicki [36], we searched manually for further literature that was missed by the automatic database inquiry. Following references of the already found papers, we looked for relevant publications that include further security design patterns that can be useful for our study.

Following several iterations, we identify a collection of papers that fulfill the inclusion criteria outlined in our SLR. Upon eliminating duplicate entries previously identified during the initial database search, we are able to incorporate an additional eleven articles into our database search results.

4.2 Results of the Search and Selection Procedures

After conducting a search and selecting papers according to the step-by-step strategy described in Sect. 4.1, we obtain the following results (see Fig. 2): After an automatic search through five scientific databases, we select 160 suitable articles. Next, titles, abstracts and content are checked against the selected inclusion and exclusion criteria. Consequently, a total of nine appropriate articles are identified. Through a manual search, an additional eleven eligible studies are found, thus augmenting the search procedure's outcome to encompass a total of 20 articles. These articles collectively represent a comprehensive compilation of 61 design patterns specifically focused on addressing security concerns in the context of the IoT (see Table 1).

Table 1. Overview of primary IoT security pattern studies.

Year	Author	Title
2021	Fernández *et al.*	A Pattern for a Secure IoT Thing [2]
2021	Papoutsakis *et al.*	Towards a Collection of Security and Privacy Patterns [19]
2020	Fernández *et al.*	Abstract and IoT security segmentation patterns [6]
2020	Fernández *et al.*	Secure Distributed Publish/Subscribe (P/S) pattern for IoT [7]
2020	Fernández *et al.*	A Pattern for a Secure Cloud-Based IoT Architecture [8]
2020	Muñoz *et al.*	TPM, a Pattern for an Architecture for Trusted Computing [14]
2020	Orellana *et al.*	A Pattern for a Secure Sensor Node [16]
2019	Moreno *et al.*	BlockBD: A Security Pattern to Incorporate Blockchain... [13]
2018	Ali *et al.*	Applying security patterns for authorization of users in IoT-based app-s [1]
2018	Schuß *et al.*	IoT Device Security the Hard(Ware) Way [27]
2018	Seitz *et al.*	Fogxy: An Architectural Pattern for Fog Computing [28]
2018	Tkaczyk *et al.*	Cataloging design patterns for internet of things artifact integration [31]
2017	Lee *et al.*	A case study in applying security design patterns for IoT... [12]
2017	Reinfurt *et al.*	Internet of Things Security Patterns [24]
2016	Sinnhofer *et al.*	Patterns to Establish a Secure Communication Channel [29]
2016	Syed *et al.*	A Pattern for Fog Computing [30]
2015	Ur-Rehman *et al.*	Secure Design Patterns for Security in Smart Metering Systems [32]
2014	Ciria *et al.*	The History-Based Authentication pattern [3]
2007	Fernández *et al.*	Security Patterns for Physical Access Control Systems [5]
2005	Kienzle *et al.*	Security patterns repository, version 1.0 [10]

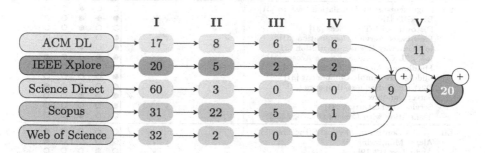

Fig. 2. Overview of the search and selection procedure results consisted of the following states: results after initial search (I), after reviewing the title and abstract (II), after scanning content (III), after cross-check information (IV) and manual search results (V).

5 Catalogue

In this section, all IoT security design patterns collected during the previously explained search process are listed in the form of a catalogue for developers.

Table 2 represents our catalogue and is divided into seven layers according to WFRM, where each pattern corresponds to a specific layer. Additionally, the possibility to solve ten vulnerabilities using each design pattern is reflected. Finally, security objectives are also mentioned that are either addressed (●) or not (○) by this particular pattern, the decisions are made based on the original description of the design pattern.

Table 2. IoT security pattern lookup table.

Layer	Pattern Name	1	2	3	4	5	6	7	8	9	10	C	I	A	Ac	Az
L1	Hardware IoT Security [27]	✓		✓	✓							●	○	○	●	○
	Secure IoT Thing [2]	✓	✓			✓						●	●	●	○	○
	Secure Sensor Node [16]			✓		✓		✓				●	●	●	●	●
	Security Segmentation [6]			✓								○	○	○	●	●
	Trusted Platform Module [14]						✓	✓				●	●	○	●	●
L2	Authenticated Channel [19]	✓				✓						○	●	○	●	○
	Encrypted Channel [19]	✓				✓						●	○	○	○	○
	Middleware Message Broker [31]			✓		✓						●	○	○	○	○
	Middleware Selfcontained Message [31]					✓						●	○	○	○	○
	Orchestration of SDN Network Elements [31]			✓		✓						●	○	○	○	○
	Outbound-Only Connection [24]						✓	✓				●	○	○	○	○
	Password-Based Key Exchange [29]	✓				✓						●	○	○	●	○
	Safe Channel [19]		✓			✓						○	●	○	○	○
	Secure Remote Readout [32]					✓						●	○	○	●	○
	Signed Message [19]					✓						○	●	○	●	○
	Symmetric Key Cryptography [29]					✓						●	○	○	○	○
	Third Party Based Authentication [29]					✓						●	○	○	●	○
	Trusted Communication Partner [24]					✓	✓					●	○	●	●	●
	Web of Trust [29]					✓						○	●	○	●	○
L3	Fog Computing [30]	✓	✓			✓						●	●	●	●	●
	Fogxy [28]	✓	✓			✓						○	○	●	●	●
	Secure Cloud-based IoT Architecture [8]			✓								●	●	●	●	●
L4	Encrypted Storage [10]						✓	✓				●	○	○	○	○
	Redundant Storage [19]					✓						○	○	●	○	○
	Safe Storage [19]						✓	✓				○	●	○	○	○
L5	Alignment-based Translation Pattern [31]			✓								○	○	●	○	○
	BlockBD [13]			✓		✓						●	●	○	○	○
	Discovery of IoT Services [31]			✓								○	○	●	○	○
	Flow-based Service Composition [31]			✓		✓						●	○	○	○	○
	IoT Gateway Event Subscription [31]			✓		✓						●	○	○	○	○
	IoT SSL Cross-Layer Secure Access [31]			✓								●	●	○	○	○
	Middleware Message Translator [31]			✓								○	○	○	○	○
	Middleware Simple Component [31]											●	○	○	○	○
	D2D REST Request/Response [31]			✓								●	○	○	○	●
	Server Sandbox [10]								✓	✓		●	●	●	○	○
	Service Orchestration [31]			✓		✓						●	○	○	○	○
	Translation with Central Ontology [31]			✓								○	○	●	○	○
L6	Access Control to Physical Structures [5]									✓		○	○	○	●	●
	Alarm Monitoring [5]							✓				○	○	○	○	●
	Audit Log [12, 19]							✓				○	●	○	○	○
	Authenticated Session [19]			✓								○	○	○	●	○
	Authorization Enforcer [19]							✓				○	○	○	○	●
	Encrypted Processing [19]							✓				●	○	○	○	○
	Fault Management [19]							✓				○	○	●	○	○
	File Authentication [1]							✓				○	○	○	●	●
	Matrix Authentication [1]							✓				○	○	○	●	●
	Minefield [10]					✓		✓				○	●	○	○	○
	Remote Authenticator/Authorizer [1]							✓				○	○	○	●	●
	Role Based Access Control [1]							✓				○	○	○	○	●
	Safe Processing [19]							✓				○	●	○	○	○
	Secure Distributed Publish/Subscribe [7]							✓				●	●	●	○	○
	Uptime [19]							✓				○	○	●	○	○
L7	Account Lockout [19]	✓										○	○	○	●	○
	Authentication Enforcer [19]		✓									○	○	○	●	○
	Blacklist [19, 24]		✓					✓				○	○	○	●	●
	History-Based Authentication [3]	✓										●	●	○	●	●
	Permission Control [24]					✓		✓	✓			●	○	○	○	●
	Personal Zone Hub [24]					✓						●	○	○	○	●
	Relays [5]											○	○	○	○	●
	Single Access Point [19]		✓									○	○	○	●	●
	Whitelist [24]		✓			✓						○	○	○	●	●

6   Discussion

To ascertain potential avenues for the expansion of our research, we conducted a thorough analysis of the compiled data regarding the patterns' capabilities in operating across various architectural levels, mitigating prevalent vulnerabilities, and addressing paramount security properties. To facilitate a more comprehensive exploration of our findings, we formulate several research questions, which were subsequently addressed and answered through a structured examination of the results.

RQ1: Which Layer in the IoT WFRM Is Covered by the Least Security Patterns? In order to answer this question, we calculated the distribution of patterns for each WFRM layer. The results are illustrated with a corresponding pie chart in Fig. 3.

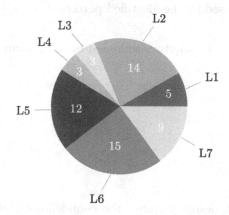

Fig. 3. Distribution of IoT security patterns among the WFRM architecture layers.

According to the calculated distribution of IoT security patterns that are attributed to the different architecture layers, most patterns can be almost equally found in the Connectivity (L2) and Application (L6) layers. On the other hand, the Edge Computing (L3) and Data Accumulation (L4) layers have only three patterns each, hence, they are on the lower end of the pattern coverage. Upon closer examination of the underlying factors contributing to this phenomenon, it can be observed that Edge Computing can be regarded as an autonomous technology infrastructure, which is not universally recognized as an integral component of the IoT across all models and frameworks. If we specifically search for Edge functionality in publications, our success rate in finding such patterns would definitely be significantly higher. But because IoT is the main focus of our research topic, only a few publications that specified IoT as well as Edge technology at the same time could be found. The underrepresentation of data accumulation (L4) in the safety patterns which we observe in our study provides an intriguing foundation for future research endeavors, warranting further in-depth analysis and investigation. The hypothesis is that the limited number of patterns currently available for addressing security issues in IoT devices can be attributed to the significant role of secure storage in both mobile and stationary computing systems. Thus far, only a few patterns have been developed with a specific focus on resolving security challenges in the realm of IoT devices. In summary, to achieve a more balanced distribution of patterns within the WFRM architecture, it is strongly encouraged to focus on the development of security patterns specifically designed for the Data Accumulation layer in the context of the IoT.

RQ2: Which Security Goals Are Covered by the Patterns? In order to address the posed inquiry, we conduct an assessment of data encompassing design patterns and security objectives, as presented in Table 1. The tabulated results, available in Table 3, illustrate the cumulative frequency at which each security goal is addressed by the identified patterns.

Table 3. Security objectives addressed by IoT security patterns.

Security Objective	Pattern Count
Confidentiality	30
Integrity	20
Availability	18
Authentication	27
Authorization	24

With a total of 30 design patterns, the confidentiality objective is the goal that is covered the most in our data set. Given its paramount importance, the protection of sensitive data is typically accorded the highest priority among various security requirements. Hence, even if an IoT system is built without any security aspects in mind, the probability that confidentiality is ensured is pretty high. Therefore, there are many patterns that guarantee this objective. Availability, however, is the security goal with the least amount of coverage. The target is to ensure that a system is accessible on user demand can be quite challenging.

Figure 4 presents the coverage of different security objectives in each layer of the WFRM architecture by design patterns in consideration.

The outlier in the bar plot is definitely the Data Accumulation (L4) layer. With only three security goals being covered and authentication and authorization being absent entirely, hence, we can assume the lack of security solutions for the storage of IoT devices. However, the reason for authentication being neglected lies in these mechanisms usually being implemented in higher layers of the IoT architecture. Interesting to mention is also the lack of availability support in the Collaboration & Processes (L7) layer. But because this layer is focused on user interactions and not the applications of the system themselves, it makes sense that availability is not a priority here.

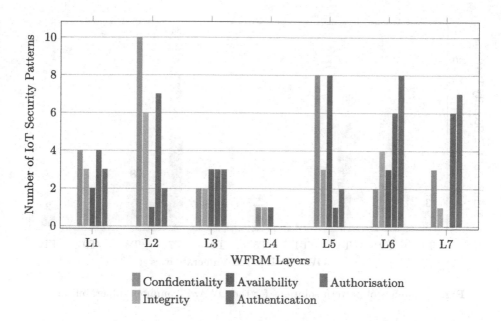

Fig. 4. IoT security patterns with addressed security concerns according to WFRM layers.

RQ3: Which Vulnerabilities from the OWASP Top Ten IoT List Are Possible to Solve by Security Patterns Included in Our Catalogue? Figure 5 shows which common vulnerabilities can be solved with the found IoT security patterns.

According to the obtained results, we can notice that each common vulnerability defined by OWASP can be solved by at least one IoT security pattern from our catalogue. The most covered is T7 which focuses on insecure data transfer and storage. At least 34 different IoT security patterns that we found have a solution to enhance the security of data handling in IoT devices. On the other hand, T4 is apparently the hardest one to solve with only one pattern addressing this issue. If we look into its description, the problem is the lack of ability to securely update the IoT device. This is a very specific issue that also is highly dependent on the hardware of the device and its user interface. For better update management of IoT devices, further research in terms of security patterns is highly recommended in this area.

Additionally, we examine the WFRM layer distribution of the IoT security patterns for each individual OWASP vulnerability. In Fig. 6, the pie charts display the more detailed results of the previous bar plot. While the pie charts for T2, T3 and T7 show the most diverse range of pattern solutions from four up to six different layers, T4 is covered by only patterns from the Physical Devices & Controllers (L1) layer.

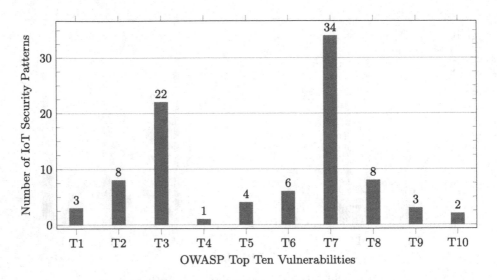

Fig. 5. Number of pattern solutions for the OWASP common vulnerabilities.

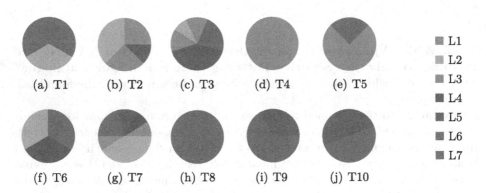

Fig. 6. Layer distribution of pattern solutions for OWASP common vulnerabilities.

Therefore, we see a connection between the found IoT security patterns for a specific OWASP vulnerability and the number of covered layers. This assumption is confirmed by the results displayed in Fig. 7. It showcases the correlation between the pattern quantity and their layer distribution with a value of 0.868. This correlation coefficient always ranges from −1 to 1 and indicates the strength of the relationship between two variables. A value between 0.7 and 1 shows a strong connection and the positive sign indicates, that more patterns as solutions for a specific vulnerability also mean a more diverse distribution of WFRM layers for these IoT patterns.

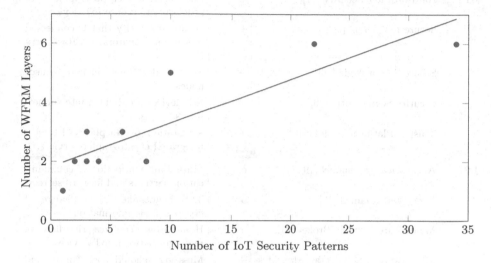

Fig. 7. The correlation between the number of patterns and WFRM layers for each OWASP vulnerability.

7   Use Case Application

In order to demonstrate a possible application of our catalogue, we have chosen the most common and at the same time vulnerable domain where IoT devices are used, specifically Smart Home. The selected scenario is simple to understand and implement, however, it encompasses most of the typical communications and activities on IoT networks that may have various vulnerabilities that need to be addressed.

Before proceeding with the detailed description, it is important to note, that in this example we assume that the system is only used by the owner of the house and no further security measures were taken than the ones that were already integrated into the system.

This smart home contains different connected devices that are distributed in the living room, bedroom, kitchen, bathroom and entrance of the house. All electronic devices run on the custom firmware Tasmota and include four RGB LED bulbs, six light controllers, and five smart plugs. Additionally, a Raspberry PI with HomeBridge allows the integration of HomeKit into the network that controls the following devices: two televisions, four Sonos ZonePlayers, a Ring camera and a doorbell.

Table 4. Use case applicability of design patterns.

Layer	Pattern Name	Rating	Explanation
L1	Hardware IoT Security [27]	+	Exchangeable cryptographic co-processors to secure IoT devices.
	Secure IoT Thing [2]	+	Secure any entity that is connected to sensors/actuators, e.g. Raspberry PI.
	Secure Sensor Node [16]	−	System does not include sensor nodes.
	Security Segmentation [6]	+	IoT devices are divided into subnetworks.
	Trusted Platform Module [14]	+	Attestation of Raspberry PI with integrated cryptographic services.
L2	Authenticated Channel [19]	o	Mutual authentication of communication partners and forward secrecy.
	Encrypted Channel [19]	o	TLS handshake and exchange of cryptographic information.
	Middleware Message Broker [31]	o	HomeBridge controls the flow of messages between IoT devices.
	Middleware Self-contained Message [31]	+	Messages should be "pure and complete" representations of events/commands.
	Orchestration of SDN Network Elements [31]	−	Only required when an IoT SDN is employed.
	Outbound-Only Connection [24]	+	Blocks incoming malicious connection requests.
	Password-Based Key Exchange [29]	+	Common secret is used to generate session key pairs.
	Safe Channel [19]	o	Use certificates to guarantee integrity during message transmission.
	Secure Remote Readout [32]	+	Security Module encrypts measurements before transmitting.
	Signed Message [19]	o	Use digital signatures during the message generation/exchange process.
	Symmetric Key Cryptography [29]	+	Handshake and common secret are exchanged between communication parties.
	Third Party Based Authentication [29]	+	Combination of asymmetric cryptography and session keys.
	Trusted Communication Partner [24]	+	List trusted communication partners and block unknown connection requests.
	Web of Trust [29]	−	Tasmota uses a central self-signed certificate authority.

Table 4. *(contniued)*

Layer	Pattern Name	Rating	Explanation
L3	Fog Computing [30]	–	No cloud-based system.
	Fogxy [28]	–	No cloud-based system.
	Secure Cloud-based IoT Architecture [8]	–	No cloud-based system.
L4	Encrypted Storage [10]	+	Critical data is encrypted before it gets committed to disk.
	Redundant Storage [19]	–	No cloud-based system.
	Safe Storage [19]	+	Guarantee integrity of stored data.
L5	Alignment-based Translation Pattern [31]	o	HomeBridge enables interoperability between different platforms.
	BlockBD [13]	–	No Big Data system.
	Discovery of IoT Services [31]	–	No usage of different IoT services.
	Flow-based Service Composition [31]	–	No usage of different IoT services.
	IoT Gateway Event Subscription [31]	o	HomeBridge sends notifications on updates.
	IoT SSL Cross-Layer Secure Access [31]	o	Only authenticated entities are able to access the external interfaces.
	Middleware Message Translator [31]	o	HomeBridge enables interoperability between different platforms.
	Middleware Simple Component [31]	+	Universally applicable pattern to achieve the best component decomposition.
	D2D REST Request/Response [31]	o	HomeBridge API is used to connect to different IoT devices.
	Server Sandbox [10]	+	Isolate server to protect it in case the system gets compromised.
	Service Orchestration [31]	–	No usage of different IoT services.
	Translation with Central Ontology [31]	o	HomeBridge enables interoperability between different platforms.

Table 4. *(contniued)*

Layer	Pattern Name	Rating	Explanation
L6	Access Control to Physical Structures [5]	–	No physical structures need to be accessed.
	Alarm Monitoring [5]	o	Alarm functionality is included in HomeBridge.
	Audit Log [12, 19]	o	HomeBridge has a rolling log screen.
	Authenticated Session [19]	–	System runs on a local server with no internet requirements.
	Authorization Enforcer [19]	–	Only relevant if a system is used by users with different roles.
	Encrypted Processing [19]	+	Integrity of data with e.g. homomorphic functions.
	Fault Management [19]	+	Smart handling of any faulty behaviour of the system.
	File Authentication [1]	–	Only relevant if a system is used by users with different privileges.
	Matrix Authentication [1]	–	Only relevant if a system is used by users with different privileges.
	Minefield [10]	+	Modify Raspberry PI to confuse attackers and simplify threat detection.
	Remote Authenticator/Authorizer [1]	–	System runs on a local server with no internet requirements.
	Role Based Access Control [1]	–	Only relevant if a system is used by users with different roles.
	Safe Processing [19]	+	Guarantee integrity during data processing with e.g. integrity checks.
	Secure Distributed Publish/Subscribe [7]	o	HomeBridge sends notifications on updates.
	Uptime [19]	o	HomeBridge measures and displays the server availability.
L7	Account Lockout [19]	o	Login via password authentication.
	Authentication Enforcer [19]	+	Authentication process that creates proof of identity.
	Blacklist [19, 24]	+	Identification of abusers who are not granted access to the system.
	History-Based Authentication [3]	+	Authentication is based on the user's own history.
	Permission Control [24]	+	User can control which data is shared with the server.
	Personal Zone Hub [24]	–	No cloud-based system.
	Relays [5]	–	No switches in the system.
	Single Access Point [19]	+	Only one entry point into the system with HomeBridge UI.
	Whitelist [24]	+	Identification of trusted partners.

After we established the use case example, we can inspect each pattern of our catalogue and check its applicability in the given context. The evaluation results can be found in Table 4 with (−) indicating this pattern is not suitable for this system, (o) being used in cases where the given system has already implemented similar security features this pattern would provide, and (+) marking recommended patterns to implement to further optimize the system design.

While there are many patterns that are not suitable to be implemented in this kind of smart home scenario, like BlockBD [13] or the Web of Trust [29], just as many are already integrated into the system, e.g. Uptime [19] or Account Lockout [19]. Nevertheless, by going through the catalogue and analysing each pattern individually, we found 25 patterns that can be used to optimize the security measures in this smart home example. Spread across all layers of the WFRM architecture, one can choose from a variety of patterns that include Symmetric Key Cryptography [29], Server Sandbox [10] or even simpler solutions like a combination of a Blacklist [19, 24] and a Whitelist [24].

This use case demonstrates that our template catalogue can serve as a simple guide to improving system security.

8 Conclusion

The aim of this analysis was to create a comprehensive catalogue of IoT security design patterns and to provide a guide for the future development of secure IoT systems.

In the beginning, we defined which IoT devices belong to the IoT spectrum. Further, we selected and described the base elements of our catalogue, such as the list of IoT WFRM architecture layers, the common IoT vulnerabilities and the most important security objectives.

In order to obtain representative results, we answer on three following questions during our research:

RQ1: Which layer in the IoT WFRM is covered by the least security patterns?
RQ2: Which security goals are covered by the patterns?
RQ3: Which vulnerabilities from the OWASP Top Ten IoT list are possible to solve by security patterns included in our catalogue?

Among the 61 design patterns in the catalogue, almost half are applied at two of the seven layers of conventional architecture. We also found a lack of coverage of security goals at the Data Accumulation level. On the other hand, every vulnerability out of 10 on the OWASP list was addressed by at least one pattern, which is a positive discovery.

A collection of security patterns in the IoT field is a good start to get an overview of the current state-of-the-art. But there are many other ways in which researchers and developers can advance secure IoT development and utilize the advantages of standardization. For future work, we identify two possibilities.

The first one is the pattern catalogue expansion. Our IoT security pattern catalogue cannot be called complete in any way. There are surely more security patterns that can be modified into the IoT context as well as other types of patterns that can make the implementation of secure IoT systems easier. A few examples would be privacy patterns, misuse patterns or anti-patterns. Therefore, the expansion of the security pattern catalogue for IoT is definitely a topic for further research.

As a second trajectory for the development of this work, we propose industry practical validation. Technology and science are ever-evolving, therefore the need for different types of patterns for common problems are always exist and require new and modern solutions. The best way to develop new security patterns, that are optimized for applicability and usage in real-world situations, is cooperation with the industry. Only when academia combines its theories and ideas with the practical problems of the corresponding industry, we are able to find the best solutions to solve common issues in the world of IoT.

References

1. Ali, I., Asif, M.: Applying security patterns for authorization of users in IoT based applications. In: 2018 International Conference on Engineering and Emerging Technologies (ICEET), February 2018, pp. 1–5 (2018)
2. Fernández, E.B., Astudillo, H., Orellana, C.: A pattern for a secure IoT thing. In: 26th European Conference on Pattern Languages of Programs, EuroPLoP 2021. Association for Computing Machinery, New York (2021). https://doi.org/10.1145/3489449.3489988
3. Ciria, J.C., Domínguez, E., Escario, I., Francés, A., Lapeña, M.J., Zapata, M.A.: The history-based authentication pattern. In: Proceedings of the 19th European Conference on Pattern Languages of Programs, EuroPLoP 2014. Association for Computing Machinery, New York (2014). https://doi.org/10.1145/2721956.2721960
4. El Hakim, A.: Internet of Things (IoT) system architecture and technologies, white paper, March 2018. https://doi.org/10.13140/RG.2.2.17046.19521
5. Fernandez, E.B., Ballesteros, J., Desouza-Doucet, A.C., Larrondo-Petrie, M.M.: Security patterns for physical access control systems. In: Barker, S., Ahn, G.J. (eds.) Data and Applications Security XXI, July 2007, vol. 4602, pp. 259–274 (2007)
6. Fernández, E., Fernandez, E., Yoshioka, N., Washizaki, H.: Abstract and IoT security segmentation patterns, January 2020
7. Fernández, E., Yoshioka, N., Washizaki, H.: Secure distributed publish/subscribe (p/s) pattern for IoT, February 2020
8. Fernández, E.B.: A pattern for a secure cloud-based IoT architecture. In: Proceedings of the 27th Conference on Pattern Languages of Programs, PLoP 2020. The Hillside Group, USA (2020)
9. Fysarakis, K., Spanoudakis, G., Petroulakis, N., Soultatos, O., Bröring, A., Marktscheffel, T.: Architectural patterns for secure IoT orchestrations. In: 2019 Global IoT Summit (GIoTS), pp. 1–6 (2019). https://doi.org/10.1109/GIOTS.2019.8766425
10. Kienzle, D.M., Elder, M.C., Tyree, D., Edwards-Hewitt, J.: Security patterns repository, version 1.0 (2006)

11. Kitchenham, B., Charters, S.: Guidelines for Performing Systematic Literature Reviews in Software Engineering, Technical Report EBSE 2007-001, Keele University and Durham University Joint Report (2007)

12. Lee, W.T., Law, P.J.: A case study in applying security design patterns for IoT software system. In: 2017 International Conference on Applied System Innovation (ICASI), May 2017, pp. 1162–1165 (2017)

13. Moreno, J., Fernandez, E.B., Fernandez-Medina, E., Serrano, M.A.: BlockBD: a security pattern to incorporate blockchain in big data ecosystems. In: Proceedings of the 24th European Conference on Pattern Languages of Programs, EuroPLop 2019. Association for Computing Machinery, New York (2019). https://doi.org/10.1145/3361149.3361166

14. Muñoz, A., Fernandez, E.B.: TPM, a pattern for an architecture for trusted computing. In: Proceedings of the European Conference on Pattern Languages of Programs 2020, EuroPLoP 2020, Association for Computing Machinery, New York (2020). https://doi.org/10.1145/3424771.3424781

15. Muccini, H., Moghaddam, M.T.: IoT architectural styles. In: Cuesta, C.E., Garlan, D., Pérez, J. (eds.) ECSA 2018. LNCS, vol. 11048, pp. 68–85. Springer, Cham (2018). https://doi.org/10.1007/978-3-030-00761-4_5

16. Orellana, C., Fernandez, E.B., Astudillo, H.: A pattern for a secure sensor node. In: Proceedings of the 27th Conference on Pattern Languages of Programs, PLoP 2020. The Hillside Group, USA (2020)

17. OWASP: IoT top 10. https://owasp.org/www-pdf-archive/OWASP-IoT-Top-10-2018-final.pdf

18. Papoutsakis, M., Fysarakis, K., Spanoudakis, G., Ioannidis, S., Koloutsou, K.: Towards a collection of security and privacy patterns. Appl. Sci. **11**, 1396 (2021). https://doi.org/10.3390/app11041396

19. Papoutsakis, M., Fysarakis, K., Spanoudakis, G., Ioannidis, S., Koloutsou, K.: Towards a collection of security and privacy patterns. Appl. Sci. **11**(4) (2021). https://www.mdpi.com/2076-3417/11/4/1396

20. Qanbari, S., et al.: IoT design patterns: computational constructs to design, build and engineer edge applications. In: 2016 IEEE First International Conference on Internet-of-Things Design and Implementation (IoTDI), pp. 277–282 (2016). https://doi.org/10.1109/IoTDI.2015.18

21. Rajmohan, T., Nguyen, P., Ferry, N.: A systematic mapping of patterns and architectures for IoT security, March 2020

22. Rajmohan, T., Nguyen, P., Ferry, N.: A decade of research on patterns and architectures for IoT security. Cybersecurity **5**, 2 (2022). https://doi.org/10.1186/s42400-021-00104-7

23. Rajmohan, T., Nguyen, P.H., Ferry, N.: Research landscape of patterns and architectures for IoT security: a systematic review. In: 2020 46th Euromicro Conference on Software Engineering and Advanced Applications (SEAA), pp. 463–470 (2020). https://doi.org/10.1109/SEAA51224.2020.00079

24. Reinfurt, L., Breitenbücher, U., Falkenthal, M., Fremantle, P., Leymann, F.: Internet of Things security patterns. In: Proceedings of the 24th Conference on Pattern Languages of Programs, PLoP 2017. The Hillside Group, USA (2017)

25. Reinfurt, L., Falkenthal, M., Breitenbücher, U., Leymann, F.: Applying IoT patterns to smart factory systems. In: Proceedings of the 11th Advanced Summer School on Service Oriented Computing, pp. 1–10. IBM Research Division (2017)

26. Samonas, S., Coss, D.: The CIA strikes back: redefining confidentiality, integrity and availability in security. J. Inf. Syst. Secur. **10**(3), 21–45 (2014)

27. Schuß, M., Iber, J., Dobaj, J., Kreiner, C., Boano, C.A., Römer, K.: IoT device security the hard(ware) way. In: Proceedings of the 23rd European Conference on Pattern Languages of Programs, EuroPLoP 2018, Association for Computing Machinery, New York (2018). https://doi.org/10.1145/3282308.3282329

28. Seitz, A., Thiele, F., Bruegge, B.: Fogxy: an architectural pattern for fog computing. In: Proceedings of the 23rd European Conference on Pattern Languages of Programs, EuroPLoP 2018. Association for Computing Machinery, New York (2018). https://doi.org/10.1145/3282308.3282342

29. Sinnhofer, A.D., Oppermann, F.J., Potzmader, K., Orthacker, C., Steger, C., Kreiner, C.: Patterns to establish a secure communication channel. In: Proceedings of the 21st European Conference on Pattern Languages of Programs, EuroPlop 2016. Association for Computing Machinery, New York (2016). https://doi.org/10.1145/3011784.3011797

30. Syed, M.H., Fernandez, E.B., Ilyas, M.: A pattern for fog computing. In: Proceedings of the 10th Travelling Conference on Pattern Languages of Programs, Viking-PLoP 2016, Association for Computing Machinery, New York (2016). https://doi.org/10.1145/3022636.3022649

31. Tkaczyk, R., et al.: Cataloging design patterns for internet of things artifact integration. In: 2018 IEEE International Conference on Communications Workshops (ICC Workshops), pp. 1–6 (2018)

32. Ur-Rehman, O., Zivic, N.: Secure design patterns for security in smart metering systems. In: 2015 IEEE European Modelling Symposium (EMS), pp. 278 283 (2015)

33. Washizaki, H., Ogata, S., Hazeyama, A., Okubo, T., Fernandez, E.B., Yoshioka, N.: Landscape of architecture and design patterns for IoT systems. IEEE IoT J. **7**(10), 10091–10101 (2020). https://doi.org/10.1109/JIOT.2020.3003528

34. Washizaki, H., et al.: Landscape of IoT patterns. In: 2019 IEEE/ACM 1st International Workshop on Software Engineering Research & Practices for the Internet of Things (SERP4IoT), pp. 57–60 (2019). https://doi.org/10.1109/SERP4IoT.2019.00017

35. Wedyan, F., Abufakher, S.: Impact of design patterns on software quality: a systematic literature review. IET Softw. **14**(1), 1–17 (2020). https://doi.org/10.1049/iet-sen.2018.5446. https://ietresearch.onlinelibrary.wiley.com/doi/abs/10.1049/iet-sen.2018.5446

36. Wohlin, C., Prikladniki, R.: Systematic literature reviews in software engineering. Inf. Softw. Technol. **55**, 919–920 (2013). https://doi.org/10.1016/j.infsof.2013.02.002

Towards a Unified Abstract Architecture to Coherently and Generically Describe Security Goals and Risks of AI Systems

Henrich C. Pöhls[✉][iD]

University of Passau, Passau, Germany
hp@sec.uni-passau.de

Abstract. We propose a unified abstract architecture for describing IT security goals and risks within AI systems. The architecture facilitates effective interdisciplinary communication among AI developers, data scientists, and security professionals. The architecture is abstract enough to cover a wide range of AI methods (not limited to machine learning) while it can still be used to sufficiently describe and map existing AI-specific attacks. It emphasises the importance of identifying at-risk processes or at-risk data within the AI system for a targeted increase of the overall system's security. This systematic approach could help to optimise resource allocation while achieving desired protection targets for AI systems.

1 Introduction

To increase the security of systems that exhibit artificial intelligence (AI) it seems rather obvious to "incorporate AI developers, data scientists, and AI-related applications and infrastructure into your security program" [11] as suggested by the Open Worldwide Application Security Project (OWASP). However, this highly interdisciplinary team then still needs to discuss the AI system, including its inner components and its interfaces to outside surrounding systems, in a coherent and exact –but still abstract enough– language. With fast moving targets like AI and information technology (IT) security these terms might not only be understood differently between different disciplines, but also within each discipline [8].

This paper provides a unified abstract architecture that can be facilitated to describe IT security protection goals, or the risks due to the absence of protection for the relevant elements (process, data, parameters) inside such an AI system. The given architecture leans towards modern machine learning terminology, but the AI system is described with sufficient generality to span wide areas of AI methods, i.e. sub-symbolic as well as symbolic, but detailed enough to identify the components targeted in existing AI-specific attacks. Of course not all processes and related elements can be found in all AI systems, i.e. logic-based or rule-based approaches are less data centred in their model generation. The paper leads the reader towards this systematic view of an AI system's architecture[1]

---

[1] We deliberately did not call the resulting architecture a 'model of the AI system' as in the world of AI 'model' is strongly reserved term.

© The Author(s), under exclusive license to Springer Nature Switzerland AG 2023
R. Rios and J. Posegga (Eds.): STM 2023, LNCS 14336, pp. 85–94, 2023.
https://doi.org/10.1007/978-3-031-47198-8_5

by presenting existing background from standards like ISO 22989 and present definitions of the terms used.

IT security targets such as integrity, confidentiality, authenticity or availability always refer to an object for which this protection target is to be achieved. Those objects requiring protection can be generalised into processes, data, or physical components. When it comes to AI specific security challenges it would be rather helpful to pin-point inside an AI system what process or data is at risk and needs protection. Pin-pointing is necessary to look specifically for the individual measures for increasing protection (so called IT security controls) for the intended protection targets. Thus ultimately, the decomposition is necessary from an economic point of view, as it helps to sensibly limit the scope of the corresponding measures and thus to use resources for a targeted and purposeful enforcement of security, safety and privacy protection targets.

1.1   Example: Poisoning Attacks Map to Integrity Attacks on Data

To exemplify, methods based on machine learning require trustworthy data for their training. In the so called 'data poisoning attack' [11] or 'poisoning attack' [5] an attacker alters training data, which also includes changing just the labels of the data, in order to have the victim train a model with attacker manipulated behaviour. The result of such an attack can be a non-functional model, i.e. an act of sabotage on the model and the AI system using it, or the attacked model might make decisions in the attacker's favour, i.e. like a Trojan horse. By the required modification of the training data the attacker violates the authenticity and integrity of the training data. Due to the failure to protect the integrity of the training data the AI system containing that malicious model exhibits undesirable and unintended functionality compared to an AI system that comprises the model generated from unmodified training data results. In other words, data poisoning attacks are integrity and authenticity attacks on training data. To mitigate 'data poisoning attacks' it thus follows that one shall:

- either take steps to ensure that integrity and authenticity are unharmed for the training data before the process of training the model,
- or "assume that the datasets used have been compromised or poisoned" [5].

In the former case classical IT security controls, technical as well as organisational ones, could help. In the latter case one could either add processes that detect and remove such data from the data sets prior to the training process or use a training processes that is less susceptible to such manipulation.

This example shows that it helps to pin risks, security goals and protection mechanisms (controls) to the affected or attacked AI component (processes or data) as it allows to map and thus communicate AI risks to IT security experts.

1.2   Terminology

In the following we briefly define the system idea that underlies the terminology of 'AI system' as well as the goal of generating an abstract system view in order to enable a risk analysis on an abstracted system level.

What Is an AI System? In the terminology standard ISO 22989 an AI system is defined as an engineered system that generates outputs such as content, forecasts, recommendations or decisions for a given set of human-defined objectives [4]. Note that our view needs to be abstract enough so that it is not geared towards specific AI methods, e.g. the architectural elements in the end shall work for sub-symbolic methods like machine learning as well as for symbolic methods. In respect to classifying AI methods we suggest the AI-MC² approach [9] which clusters into: "Problem Solving, Optimizing, Planning and Decision Making"; "Knowledge Representation and Reasoning"; "Machine Learning"; "Hybrid Learning". For an abstract architecture we need to further dissect the AI system, thus we see an AI system more generically as a system that uses artificial intelligence and consists of several components (at least one of which is an AI component). The AI component then is the component that involves AI methods and/or provides AI functionality [3,8].² Especially, our goal is to split the AI component into several sub-components. We will use the following terms:

- **AI system** is an engineered system that generates outputs such as content, forecasts, recommendations or decisions for a given set of human-defined objectives. The AI system consists of several components at least one of which is an AI component.
- **AI component** is a system component that provides artificial intelligence; consisting of several sub-components.

What Is the AI System Life-Cycle? There are several phases within any system's life [6]. Likewise the AI terminology standard 22989 from ISO defines that there are the eight stages "inception", "design and development", "verification and validation", "deployment", "operation" and "monitoring", "re-evaluate", "retirement" [4]. This is, of course, not generally different for AI systems. From a security point of view it is important when protection needs to be applied, i.e. in which stage of the life cycle an attack would happen. For our abstract architecture we will thus identify which of the processes or datasets for the AI component are involved in which life-cycle stage (see Sect. 3.3).

2 Existing Component Diagrams of AI Systems

In the following we will take a look at existing decomposition of AI systems. Note, not all are supposed to model the whole AI system as we defined it, nor are all geared towards AI method independence. However, they were taken into account when deriving the proposed architecture (see Sect. 3). We start with the one from ISO 22989, which is the international standard for AI terminology from 2022 (see Fig. 1). Then we look at the one from the International Software Testing Qualifications Board (ISTQB) which is more workflow and life-cycle

---

² As there is too much debate on what suffices as a concrete definition of artificial intelligence, we leave this discussion aside in this paper.

phase oriented. As we are interested in arriving at an architecture that can be used to describe existing and future risks and safeguards for an increased security of AI systems we look at the architectures of AI systems used by the Berryville Institute of Machine Learning (BIML) [7] from Gary McGraw who is also a specialist on secure software [12] and their usage for risk assessment. We finally look at the AI system described by the Open Worldwide Application Security Project (OWASP) when describing AI security and privacy [11].

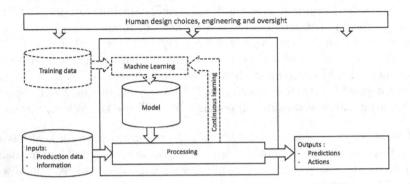

Fig. 1. Components in the functional view according to ISO 22989 [4]

Figure 1 from ISO 22989 [4] shows a machine learning centred view with the model in the centre that is received by training it using machine learning based on the training data. The model is then used in a process called "processing" to generate outputs (i.e. predictions or actions) from inputs. It also features a circular arrow to show that there might be AI components which are continuously learning. While this view is correct it does not capture a process that could be used to anchor how training data shall be generated, e.g. to avoid AI specific risks like over-fitting or bias. Figure 1 also fails to highlight that there needs to be evaluations and tests in the form of additional processes with different datasets. While this can all be subsumed into the box with "Human design choices, engineering and oversight," [4] this is not very detailed yet.

Figure 2 from ISTQB [3] is again tailored to provide a view of machine learning. Thus, it is only geared towards one class of AI methods [9]. Unlike the ISO architecture diagram, it puts more emphasis on the different phases, i.e. "Deploy the model" then "use the model" [3]. From a security point of view it clearly adds the need for testing the model, which is also the focus of the International Software Testing Qualifications Board (ISTQB).

Figure 3 from BIML [7] is also for machine learning only. It clearly identifies the need for three different datasets ("training, validation, test" [7]) and nicely highlights that the separate process that assembles these datasets needs careful design. They map the risk of data poisoning[3] as follows to their sub-components:

---

[3] BIML has a nice interactive online version: https://berryvilleiml.com/interactive/#data [last accessed: Jun. 2023].

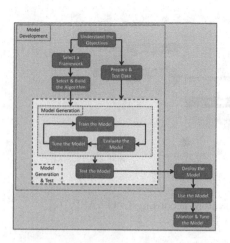
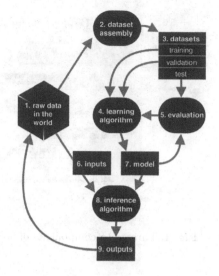

Fig. 2. Components of an AI system in the machine learning workflow model from ISTQB [3]

Fig. 3. Components of an AI system associated with risks from BIML [7]

"... raw data in the world, dataset assembly, and datasets ... are subject to poisoning attacks whereby an attacker intentionally manipulates data in any or all of the three first components, possibly in a coordinated fashion, to cause ML training to go awry." [7] They have used shapes to differentiate three classes of sub-components: processes (round), data (squared) and raw data (diamond). They have shown that their components are able to be used to describe existing risks and have mappings for 78 specific risks associated with their view of a generic AI system using machine learning.

Finally, Fig. 4 from OWASP [11] also states the threat of "Data poisoning" and maps it to their section of "Data Engineering" and the "Source data" therein. It takes a more software development view and shows that the AI-related or "data science model attack prevention" is only one surface that needs to be protected against attacks. They take an initial system of systems view when they propose to "throttle & monitor" some parts of the system.

3 Proposed Abstract Architecture of an AI System

The proposed abstract architecture (Fig. 5) takes into account sub-components and facets that worked specifically well for the existing 'frameworks' discussed before. This includes a system of systems approach and life cycle phases.

3.1 Proposed Architecture Covers Not only Machine Learning

Note that our architecture is heavily influenced by terminology used in machine learning (ML), and also has many processes that on first hand look like they can

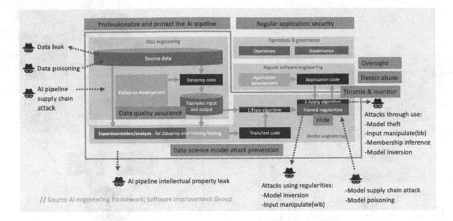

Fig. 4. Threats to components of an AI system according to OWASP [11]

only be applied to ML, but they are meant generically. To give an example, if you generate a knowledge base by interviewing human experts to build a decision tree, then this process of interviewing would be the 'model generation' process and your encoded decision tree would be the 'model'.

3.2   System of Systems Approach of the Proposed Architecture

We view the AI system as a generic system that takes input (upper left corner of Fig. 5) and produces outputs (lower right corner of Fig. 5). The AI system then comprises several sub-systems of which one is the AI component. This is at the centre of our attention to describe AI specific risks and their mitigation, but of course the AI component shall not be considered in vein. Thus, we have highlighted that other AI system components could be influencing input and output as well as supervising the AI component. This could be a positive or negative influence, i.e. it could be that these components increase the security of the system by running additional checks on inputs or outputs, or decrease the security because they might strip away useful details, i.e. remove the information on the confidence of a classifier before processing the result in upstream systems. Also, the software implementing AI sub-components itself is subject to software security vulnerabilities [2].

3.3   Life Cycle Phases of the Proposed Architecture

We identify five crucial phases, which are easy to differentiate and to communicate across interdisciplinary teams. They are colour-coded and ordered by time, i.e. processes that finish before others are placed closer to the top. We foresee them happening in this order: design, modelling, validation, testing, deployment. Of course one could go back and repeat, e.g. do a re-modelling if the testing failed.

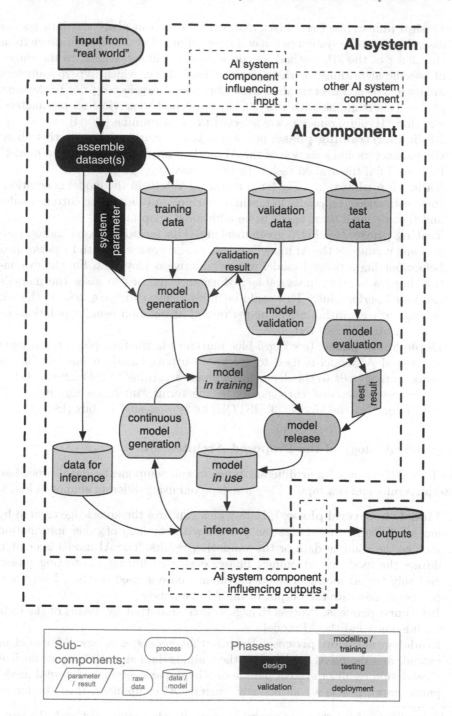

Fig. 5. Proposed sub-components of the abstract architecture of an AI system's AI component (Color figure online)

- **Design phase:** (see Fig. 5 black marking) During this phase the datasets are assembled and the system parameters are defined. Parameters can range from the choice of the ML method, i.e. do we use machine-learning, to the choice of specific algorithms, or their parametrisation, i.e. which hyperparameters are used. The architecture shall show that these are selected prior to the next phase. This is in-line with ISO 22989 stages and their view on parameters, e.g. that "Hyperparameters are selected prior to training [..]" [4].
- **Modelling/training phase:** (see Fig. 5 yellow marking) During this phase the parameters and selection of the AI method are used to generate a model[4] that can fulfil the desired task with the desired level of quality.
- **Validation phase:** (see Fig. 5 red marking) Whilst in the model generation, there is the need, especially in machine learning to validate the current model and then define if there needs to be additional adaptations.[5]
- **Testing phase:** (see Fig. 5 green marking) This includes the evaluation process and it controls the AI model's release, i.e. there are tests to be performed before putting a trained model into productive deployment. So, the evaluation process is using measurable performance metrics to asses the model's functional performance. This could be quite a huge effort, e.g. testing the self driving car's algorithm might be performed in isolation using simulations or in a real car on a real test drive.
- **Deployment phase:** (see Fig. 5 blue marking) In the final phase the trained and tested AI model is used to generate outputs based on inputs. This of course corresponds to the deployment phase according to ISO 22989. Herein the process 'inference' corresponds to the terms "inference algorithm" [7] (BIML) or "Use the Model" [3] (ISTQB) or "Processing" [4] box (ISO 22989).

3.4 Terminology of the Proposed Architecture

We have differentiated four different types of sub-components: data, processes, interim results and raw input. They are depicted using different shapes in Fig. 5.

- **Model (in several phases):** Data which captures the knowledge created by means of the AI model creation process with the help of other information such as the training data or the validation results. The AI model is created during the modelling (training) phase, evaluated during the testing phase, and only the AI model in the deployment phase is used by the AI inference process to generate the output of the AI component.
- **Inference process:** Process that generates an output by means of the data for inference and the AI model.
- **Model generation process:** Process that generates a new AI model or extends an existing AI model using the training data and other inputs such as a validation result or system parameter (hyperparameters, or internal model parameters). If it is a continuously learning AI component, then the model

---

[4] We explicitly added "training phase" as many people familiar just with the latest trend of machine learning would understand that this phase generates the model.

[5] We named this process 'model validation'; it is also referred to as model tuning.

generation process also takes information from the inference process (results, internal values, but also data for the processing) as additional inputs.

- **Model validation process**: Process that uses validation data and other parameters to evaluate an existing AI model and drive the AI model generation process.
- **Model evaluation process (within the context of the test process)**: Process that evaluates an existing AI model using the evaluation data and produces a test result.
- **Assemble dataset(s) process**: Process for transforming the input (raw data) into a representation such that it becomes suitable for the following processes (model generation process, inference process, model validation process, model evaluation process) of the AI component.
- **Training data**: Data for the model generation process, generated before by a data curation run in the assemble datasets process.
- **Validation data**: Data for the internal model validation of the training process, generated before by a data curation run in the assemble datasets process.
- **Test data**: Data for the model evaluation process, generated before by a data curation run in the assemble datasets process.
- **Input (raw data)**: Unprocessed data passed to the AI component as input from the AI system, for example, image content in the case of an image recognition AI component.
- **Data for inference**: Data generated before by a data curation run in the assemble datasets process that is fed to the AI inference process to produce an output.
- **Output (data)**: Results of the AI component, which were determined by applying an AI algorithm in the inference process using an AI model to the input.

4   Conclusion

In summary, our proposed architecture is on one hand abstract enough to cover different AI methods, including but not limited to machine learning[6]. Only exemplary, due to the limited space, we showed that our generalisations could also describe an expert system based on a decision tree. Whilst such a generalised architecture might seem overkill for some AI methods, we still see great value in being able to apply it throughout different AI methods in order to re-consider security risks already identified for other AI methods and —if not specific— also reuse mitigation methods. This would especially hold true if the risk mitigation is done using general IT security controls, such as integrity or authenticity protection using digital signatures or organisational safeguards like access control.

On the other hand, the depth of granularity is high enough to identify roles and processes, e.g. for certified AI testers, and to pin-point where security risks surface and thus security goals need to be fulfilled. This enables existing attacks

---

[6] By 'AI methods' we refer to sub-symbolic, symbolic as well as hybrid methods; see [9].

to be mapped to one architecture. Due to space limitations we could not elaborate here, but as we have a super set of the sub-components from the BIML model we are sure that their existing 78 risks easily translate into ours. Furthermore, we embedded our architecture into a birds-eye view: The AI component is within a system of systems. By this it enables describing security problems also at the boundaries of the AI component.

For the future we want to apply this work to facilitate interdisciplinary stakeholder discussions, including the communication with users of AI-enabled devices and services, as those become more and more embedded into our everyday digital life. Finally, we hope it aids to develop a harmonised view on the many recent and ongoing works on threats to AI systems; like the upcoming ISO 27090 [5], or NIST's AI 100-1 [10], or ENISA's Multilayer Framework for Good Cybersecurity Practices for AI [1], to mention only a few.

Acknowledgements. This work was partly funded by the Bavarian Ministry of Science and Arts under the ForDaySec (https://fordaysec.de/) project.

References

1. ENISA, Polemi, N., Moulinos, K., Praça, I., Adamczyk, M.: A multilayer framework for good cybersecurity practices for AI. ENISA (2023)
2. Harzevili, N.S., Shin, J., Wang, J., Wang, S., Nagappan, N.: Characterizing and understanding software security vulnerabilities in machine learning libraries. In: IEEE/ACM International Conference on Mining Software Repositories (MSR), pp. 27–38 (2023)
3. International Software Testing Qualifications Board: Certified Tester AI Testing (CT-AI) Syllabus (Version 1.0). ISTQB (2021)
4. ISO/IEC: ISO/IEC 22989:2022: Information technology - Artificial intelligence - Artificial intelligence concepts and terminology. ISO Geneva, Switzerland (2022)
5. ISO/IEC: ISO/IEC 27090:draft: Cybersecurity - Artificial Intelligence - Guidance for addressing security threats and failures in artificial intelligence systems. ISO Geneva, Switzerland (2023)
6. ISO/IEC/IEEE: ISO/IEC/IEEE 15288:2023: Systems and software engineering - System life cycle processes. ISO Geneva, Switzerland (2023)
7. McGraw, G., Figueroa, H., Shepardson, V., Bonett, R.: An Architectural Risk Analysis of Machine Learning Systems: Toward More Secure Machine Learning (2023). https://berryvilleiml.com/results/ara.pdf. Accessed June 2023
8. Runze, G., et al.: AI glossary (version 1.0) (2023). https://www.ai-glossary.org/. Accessed June 2023
9. Schmid, T., Hildesheim, W., Holoyad, T., Schumacher, K.: The AI methods, capabilities and criticality grid. KI - Künstliche Intelligenz **35**(3), 425–440 (2021)
10. Tabassi, E.: Artificial Intelligence Risk Management Framework (AI RMF 1.0) (2023)
11. van der Veer, R., et al.: AI security and privacy guide information (version 1.0) (2023). https://owasp.org/www-project-ai-security-and-privacy-guide/. Accessed June 2023
12. Viega, J., McGraw, G.: Building Secure Software: How to Avoid Security Problems the Right Way. Addison-Wesley, Reading (2001)

Decentralized Global Trust Registry Platform for Trust Discovery and Verification of e-Health Credentials Using TRAIN: COVID-19 Certificate Use Case

Isaac Henderson Johnson Jeyakumar[1]($\boxtimes$) [iD], John Walker[2]($\boxtimes$) [iD],
and Heiko Roßnagel[3] [iD]

[1] University of Stuttgart IAT, Allmandring 35, 70569 Stuttgart, Germany
isaac-henderson.johnson-jeyakumar@iat.uni-stuttgart.de
[2] SemanticClarity, 44 Bonnie Lane, CA 94708 Berkeley, USA
jwalker@semanticclarity.com
[3] Fraunhofer IAO, Nobelstrasse 12, 70569 Stuttgart, Germany
Heiko.Rossnagel@iao.fraunhofer.de

Abstract. The global COVID-19 pandemic has shown us the requisite efficacy of e-Health Credentials (COVID-19 vaccine certificates), including how they can be created quickly in compliance with different technical requirements of defined safety standards. But things get more complicated as soon as those certificates leave their state or country. A global pandemic requires e-health certificates that are valid globally so that a person's health status can be demonstrated and verified reliably across national borders. To issue and validate the certificates globally, a secured Trust Registry platform is needed by which one can discover and verify the trustworthiness of e-Health credentials. This paper proposes a new decentralized trust registry platform which can be used to issue and validate e-Health certificates on a global scale, an approach based on Domain Name System (DNS) using TRAIN (**TR**ust m**A**nagement **IN**frastructure). Decentralized trust registry platform also facilitates individual trust decisions through the discovery of trust lists so that verifiers can perform informed trust decisions about issued e-Health credentials. The proposed decentralized trust registry platform also helps to integrate existing trust frameworks into the platform so that they can be accessed globally. This paper also provides a demonstration on how this trust registry platform could be used for trust discovery and verification of COVID-19 certificates. The Trust registry platform is not restricted to specific identity ecosystems, rather it can support both legacy PKI and SSI Identity ecosystems.

Keywords: Decentralized Trust Registry · TRAIN · trust-list · e-Health Credentials · Digital COVID-19 certificates

R. Rios and J. Posegga (Eds.): STM 2023, LNCS 14336, pp. 95–104, 2023.
https://doi.org/10.1007/978-3-031-47198-8_6

1 Introduction

The challenge of providing accessible, trusted, interpretable, digital health service information and health service specific credentials and certificates at a regional or international scale is a challenge that governments, regional and international health organizations, and private sector consortia faced in the recent COVID-19 pandemic.

The COVID-19 pandemic created widespread constraints on individuals engaged in cross-border travel or participating in many types of domestic activities based on the individual's ability to demonstrate their health or immunization status. To address and minimize these constraints many governments and public and private sector health consortia worldwide responded by creating COVID-19 vaccination and test certificates [1]. These certificates are digital or paper documents that allow an individual to present their health status with respect to COVID testing or vaccination as required by their circumstance.

To support the implementation of a platform for the discovery, evaluation and verification of such certificates, this paper describes an innovative planned approach for a globally accessible multi-tenant trust registry platform. The platform leverages the Internet Domain Name System (DNS) and its security extensions and TRAIN (**TR**ust m**A**nagement **IN**frastructure) a lightweight trust infrastructure, backed with defined roles for operational governance.

A project vehicle to demonstrate this platform was initiated at Linux Foundation Public Health (LFPH) in June of 2021, the Global COVID Certificate Network (GCCN) project. The project's core goal was facilitating the safe and free movement of individuals globally during the COVID pandemic [2]. The initial project activities focus on building and demonstrating a GCCN Trust Registry Network instance. This multi-stakeholder network would provide a technology agnostic, publicly accessible mechanism for COVID certificate issuers and verifiers to:

- Publish relevant entity definition meta-data
- Discover Covid-19 Certificate issuers and certificate definitions
- Review published certificate policies
- Build a list of trusted certificate issuers and access their public keys for certificate verifications

Establishing an infrastructure for the GCCN that supports the features above, is itself widely available, and leverages existing internet technology were drivers for the adoption of TRAIN as the mechanism for trust scheme discovery and definition.

This paper contributes to establishing the GCCN by calling out challenges that need to be addressed for the delivery of a globally accessible certificate network, framing these against current efforts, and providing a candidate solution for the trust registry and its service definitions. These challenges are highlighted in Sect. 2 of the paper. Section 3 presents existing initiatives carried out toward global implementation and verification of COVID certificates. Following in Sects. 4 and 5 the architecture, registry enrollment, discovery process, and integration are discussed. Lastly, in Sect. 6 we discuss the current limitations of this architecture, along with a conclusion.

2 Trust and Interoperable Challenges in Global Covid Certificate Ecosystem

The value of a Trust Registry is demonstrated in the availability, accuracy and quality of the entities and information it represents. The primary challenge of establishing "institutional trust" for an entity [1], in the world of digital identity still requires the vetting of a candidate entity's identifying attributes by an agreed and trusted agent. While standards based best practices exist for establishing an identity [2, 3], the implementations of digital identity registries are largely siloed and without common mechanisms for vetted cross reference and common definitions. This is also the case with the COVID-19 Certificate 'registries', and for each of the prominent certificate issuer networks listed, the digital identities of issuers and their roots of trust for the certificates issued are only valid within these closed systems.

EU DCCG (European Union Digital Covid Certificate Gateway): This network [4] representing 45+ countries (27 member states and at least 15 outside states or territories) has generated more than 650 million EU Digital Covid Certificates, participation requires issuers to adhere to the EU DCC specifications which includes the stipulation that both issuers and verifiers be listed within the Gateway's public key PEM files. There are no discovery or interoperability mechanisms for issuers outside of the DCCG members.

ICAO (International Civil Aviation Organization) Network: This international advisory organization has a network with 145 participating nations, the ICAO has created its own ePassports and Visible Digital Seals (VDS-NC) [5] formats for enabling cross border travel. All issuance parties are within the ICAO's network and Public Key Directory, and verifiers must recognize ICAO's proprietary certificate format. While international in its membership, it is ultimately a siloed system.

VCI Directory: VCI [6] is a public and private sector coalition with more than 300 members, supporting issuers of COVID vaccination and test certificates in a SMART Health Card format, The VCI directory references member URL's and an unsigned directory of member's public key IDs. The root of trust for listed members is their domain URL.

DIVOC (Digital Infrastructure for Verifiable Open Credentialing): Digital Infrastructure for Vaccination Open Credentialing [7] is an open source software stack for the generation of digital and paper Covid vaccination certificates. It is embedded in nation state level COVID-19 Vaccine Intelligence Network (CoWin) portal implementations. Instigated in India, this network has issued 2 billion vaccine certificates across 5 countries does not provide a master list of members. Issuers and verifiers work with each other to establish bilateral trust and determine the roots of trust and format specifications. Generally, members expose their public keys directly to a domain. Issuer metadata, key usage, and policies vary by implementation.

For each of these networks/consortia, the definition, collection, and distribution of content and metadata necessary to support their version of a Covid-19 certificate must be performed by network participants and/or the member organization itself with no straightforward way of interoperating with the other network's formats, technologies, or policies.

3 Related Works

There are initiatives whose work is actively addressing the challenges of siloed digital health certificates networks. The World Health Organization (WHO) has published two technical specifications documents for Digital Documentation of COVID-19 Certificates: Vaccination Status, and Test Results [8]. These technical specifications and guidance documents are designed to guide countries and technologists on how to develop or adopt digital systems in support of verifiable proof of vaccination and test results for domestic and cross-border purposes. The technical specifications and implementation guidance details the use of interoperable standards, facilitated by a common digital architecture, for digitized certificate issuance.

Another project that addresses the verification of multiple known Covid certificate formats is PathCheck's Universal Verifier App [9], this application supports resolving verification conversions for DIVOC, SHC, HC1 and VDS certificate and QR code formats on both the iOS and Android mobile platforms. While demonstrating conversions of multiple certificate formats the application cannot provide further trust verification of an issuer beyond their published certificate public keys.

The following are proposed solutions for the creation and management of Digital Vaccine Certificates (DVC) using blockchain.

The authors Zhao et al. addresses [10] how smart contract based blockchain technologies could be realized in 3 scenarios of DVC. First regarding control and transmission of identity information related to Digital Vaccine Certificates. And then it addresses how blockchain technology could provide traceable and tamper-proof data storage. And last how it can be applied cross-domain for user digital identity authentication using DIDs.

The authors Pericas-Gornales et al. [11] present a blockchain-based protocol for COVID-19 digital certificate management system using a proxy re-encryption service, which provides high privacy, authenticity, and self-sovereignty of data. It also addresses the use of the IPFS distributed storage system to store the digital COVID-19 certificates, encrypted by the PRE service, which provides permanent access to all certificates, and a regulatory authority.

The authors Nabil et al. [12] address the how Ethereum based blockchain architecture called "Digital Vaccine Passport" can play a role in providing privacy friendly, transparent, and authentic process for vaccine certificate issuance and verification. The paper also describes different entities and their roles involved during the life cycle of the vaccine certificate. The paper also provided a cost efficiency and benchmark result analysis for different transactions involved in the ecosystem.

All the above blockchain approaches provide architectures and infrastructures for issuing and verifying Digital Vaccine Certificates in a secured, transparent and privacy friendly means. But interoperability between the different blockchain infrastructures, integrations of other existing PKI infrastructures, trust registry data models and validating the institutional trust of the Digital Vaccine Certificates was not analyzed or reported. The analysis of different covid certificate infrastructures described above with respect to interoperability, credential formats, trust lists and technology used has been described in the following Table 1.

Table 1. Analysis of different covid certificate Infrastructures

COVID-19 Vaccination Certificate Infrastructure	Credential Format Supported		Trust List nature				Supported Technology		Interoperable
	JSON based Proprietary	W3C Verifiable Credential	Centralized	Federated	Decentralized/ Smart Contracts	Not Used	Block Chain	PKI	
Zhao et.al	x					x	x		
Pericas-Gornales et.al	n.d				x			x	
Nabil et.al	n.d					x	x		
DIVOC		x	x					x	
EU-DCCG	x			x				x	
ICAO	x		x					x	
GCCN	x	x			x		x	x	x

4 Enabling Decentralized Global Trust Registry for Cross Domain Trust Using TRAIN

TRAIN (TRust mAnagement INfrastructure) was a subproject run by Fraunhofer-Gesellschaft in the EU NGI eSSIF-Lab initiative [13]. The conceptual approach of TRAIN as a lightweight trust infrastructure was first published in [14] which describes how TRAIN uses the global, well-established, and trusted infrastructure of the Internet Domain Name System DNS as its root of trust, leveraging DNS's ubiquitous use and recognition. These security extensions to DNS (DNSSEC) [16] have been specified to ensure that the results returned from DNS queries are authentic and have not been tampered with. Consequently, TRAIN uses DNSSEC whenever this is available. The basic technology used by TRAIN has already been developed and validated in several pilots of the EU LIGHTest project [15]. TRAIN was also integrated into the eSSIF-Lab Interoperable project to verify the trust of Verifiable Credential (VC) issuers in the SSI ecosystem and it has been published in [16]. The latest draft of OpenID Connect for Verifiable Presentations [17, 18] contains informative implementation guidelines describing how issuers, holders and verifiers can utilize the TRAIN trust scheme approach. The main goal of this paper is to articulate a technical architecture that enables a global decentralized trust registry with cross domain interoperability.

4.1 Architecture of Global Covid Certificate Network Trust Registry

This section discusses the Architecture of Global Covid Certificate Network Trust Registry as shown in Fig. 1. It also explains in detail regarding the interaction and functionalities of different components in the architecture.

Trust Scheme Publication Authority (TSPA): TSPA is responsible for creating their own trust scheme and operating the Trust Registry. As per our Global Covid Certificate Network use case scenario as shown in Fig. 1, the TSPA is responsible for enrolling the member nations into the Trust Registry and is also responsible for anchoring the PTR records of member nations to the DNS Zone Manager. Member nations around the globe who wish to join GCCN can approach the GCCN with their corresponding trust scheme details and the public key of their corresponding individual trust list and become enrolled in the GCCN trust registry. TSPA has also the possibility to delegate its Roles, For example: Different entities could delegate responsibility for enrolling continent member nations into a common trust scheme authored by their TSPA.

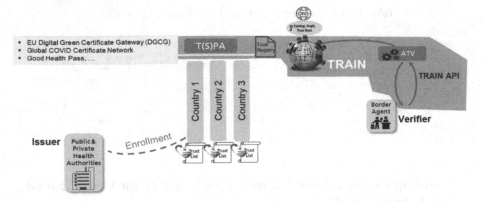

Fig. 1. Architecture of Global Covid Certificate Network (GCCN)

Individual TPA: A Trust Publication Authority (TPA) is operated by individual member nations under their domain. The TPA is responsible for onboarding the authorized issuers of their country or state into the Trust List. The Trust list is stored and operated on the premises of member nations and the member nation are flexible to decide what environment they would like to host the trust list. The Trust Scheme pointer of the TPA should point to the GCCN (TSPA).

Trust Registry: The Trust Registry contains the detailed information of individual member nations who are participants in the GCCN. Trust Registries provide the pointers to their individual trust lists. Although the Trust Registry pointer is stored in the DNS Zone Manager hosted by GCCN, the Trust Registry itself is hosted in an IPFS decentralized environment thereby preventing a single point of failure. Depending upon the usage or traffic of the infrastructures the nodes of the IPFS will be increased.

Trust List: The Trust List of individual member nations contains trusted private and public authorities who are responsible for issuing covid certificates in the respective region. The TPA is responsible for creating the onboarding process and it can vary depending on the nation either manually or automatically via an API request. The trust list is designed on the basis of an ETSI Trust list format, this trust list format contains information about cryptographic signatures (e.g. EU DCC Gateway public keys) used to verify certificates and legal information regarding the immunization service provided. The legal information represents the qualifier details of the issuing authority, at what scope are they qualified to issue the covid certificate, and the policies followed by the issuing authority.

Issuer: The issuers in this context are private and public authorities responsible for issuing covid certificates. The public keys of issuing authorities with Key ID will be enrolled with TPA during the enrollment process. The enrollment process can vary depending on the legislation where the issuers are located.

Verifier: Verifiers are entities that are responsible to validate the covid certificates issued by different authorities under different legislations globally. And trust lists of member nations play a major role in providing the verifier with the necessary information to validate the covid certificates. How a TRAIN Verifier is used to validate existing covid certificates will be discussed in Sect. 5.

4.2 Specifications for Trust Registry and Trust List

As mentioned in the previous section, the TSPA and Individual TPA are responsible for enrolling member nations and certificate issuers in the Trust Registry and Trust List. In this section specifications of the Trust Registry, Trust List and their enrollment procedures will be discussed in detail. In TRAIN both Trust Registry and Trust List follow the ETSI standard TS 119 612 [3] which is xml based. Basically, for Trust Registry Enrollment the member nations have to provide the details of the entity operating the TPA along with its corresponding scheme name and the pointer to where the local trust list is stored, so that it can be publicly available in the registry. An example of member nation Trust List can be found in the following link [19]. The data model of the trust registry and trust list includes the following mandatory information - *Identifier Details (DID/URI/UUID), Trust Scheme Details, A legal name of the service provider, An associated URL of the website, Associated e-mail address, An associated Trade Name, An Postal Address of the entity, An TSP Qualifier List, TSPServiceName, TSPService Digital Identity, x.509 Certificate, DID, ServiceGovernanceURI, ServiceHistory, ServiceStatus, ServiceTypeIdentifier.*

A *TSPService* consists of a public definition of the different services offered by a service provider issuing related certificates. This information endpoint has all information pertaining to the service along with their policies, public keys, status, and other scheme information. This trust list can be extended for different services offered by the service provider. Currently, the focus is covid certificates and the information related to covid certificates per each service. These trust lists also accommodate schema information required to verify certificates which are issued based on W3C Verifiable Credential format. These properties represent an example of the specifications required to enroll in the Trust Registry and Trust List, additionally the governance body who is operating the TSPA, or individual TPAs may require other sorts of identity proofs depending on member nation legislation and compliance requirements. These requirements, of course, can vary from one member nation to another.

4.3 Trust Discovery Process

Global Trust Discovery is one of the unique features of the TRAIN. In this section the trust discovery of covid certificates issued by different service providers around the world will be discussed in detail. A pictorial overview representing the Trust discovery approach is shown in Fig. 2. As discussed in the previous chapters regarding two-step enrollment process using TSPA and TPA, one for the enrollment of member nations and other for service providers in member nations. Similarly, the discovery of a service also involves a two step process which follows a top down approach without disturbing the chain of trust. This 2 step process is enabled via 2 APIs, the first one uses the top level scheme (*gccn.lfph.com*) to find participating member nations, and the second API is uses the results of first API to discover the trust service provider of the member nations trust list(*covid.country1.com*). Once accessed, the service provider detail API exposes the wide range of legal and cryptographic information to identify the trustworthiness of a service provider. The *ServiceDigitalID* contains the key information such as Kid, Certificates etc. The *ServiceGovernanceURI* tells us the nature of Governance which they use including the Level of Assurances (LoA). *In ServiceSupplyPoint* the scope and policies of the service provider can be defined, since it's a global infrastructure each and every member nation might not use the terms and policies. By which this information can help the verifier to access and verify the trust of different serviceproviders.

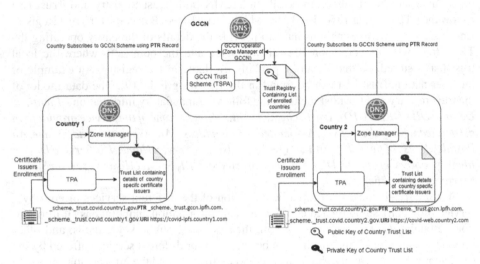

Fig. 2. Trust Discovery Approach

5 Integration and Validation of Existing COVID Certificate Platforms with the TRAIN Trust Registry

TRAIN achieves interoperability and global discovery not by creating some new ecosystem for member nations but rather by using their current infrastructure with very minimum transition overhead. The member nations do not need to reissue covid certificates using the new infrastructure but rather TRAIN gives the possibility to integrate their existing covid platform with GCCN Network. For example, the public key information of different service providers is crucial to verify the issued covid certificates. And this public key information of the existing platforms can be brought into the Trust List of the member nation under the attribute *ServiceDigitalIdentity*. This attribute can contain a URL which points to the public key information. Similarly during issuance process of certificate *trust_scheme_pointer* must be embedded as attribute in the certificate. And during validation process the verifier component of the TRAIN provides a means to validate the trust of the issued covid certificates around the world with minimum data. In the following link (https://s.fhg.de/Integration-Validation-GCCN) the integration and validation of covid certificates issued by existing ecosystem such as EU-DCC, ICAO Digital Seals and DIVOC will be discussed in detail with example data model of the certificate formats.

6 Conclusion

Decentralization is achieved using DNS by trusting DNS as the Root of Trust. Due to the distribution of trust lists, the member countries have more administrative overhead to maintain the infrastructure of Zone Manger and TPAs rather than depending on the centralized Gateway. In addition, member nations who are planning to publish the Trust Lists in IPFS (Interplanetary File System) must determine themselves the number of nodes to deploy. And while TRAIN provides for the member nation's TPA to execute or delegate the provider enrollment tasks, the enrollment process must be monitored carefully.

This paper explores in detail the architecture of decentralized global trust registry platform leveraging DNS as the backbone root of trust. It also explains the different components involved in this network and the distributed trust list enrollment process which provides member nations more autonomous control over the growth, lifecycle and infrastructure of their trust lists compared to a centralized gateway. The paper also demonstrates the interoperable nature of the trust lists, available in machine readable form, that contain a Trust Service Provider's attested and public keys, Entity Business and Policy metadata are uniquely scoped to TRAIN. It shows how the existing covid certificate platforms like EU-DCCG, ICAO, DIVOC can be integrated into this trust network without much overhead and how existing issued certificates can be validated globally. This trust management infrastructure is not restricted to COVID Certificates but can be used in different applications for trust verification of Issuers, Verifiers and Holders in both legacy PKI and SSI Identity ecosystems.

References

1. GS1: Global Legal Entity Identifier (LEI) Service. https://www.lei.direct/de/glossar/local-operating-unit-lou/. Accessed 19 Aug 2022
2. Grassi, P.A., et al.: Digital Identity Guidelines - Enrollment and Identity Proofing. https://doi.org/10.6028/NIST.SP.800-63a
3. ETSI: Electronic Signatures and Infrastructures (ESI): General Policy Requirements for Trust Service Providers (2016). http://www.etsi.org/deliver/etsi_en/319400_319499/319401/02.01.01_60/en_319401v020101p.pdf
4. EU-DCCG: eHealth Network Guidelines on Technical Specifications for EU Digital COVID Certificates Volume 2 (2021). https://ec.europa.eu/health/sites/default/files/ehealth/docs/digital-green-certificates_v2_en.pdf
5. ICAO: Visible Digital Seals ("VDS-NC") for TravelRelated Public Health Proofs. https://www.icao.int/vdsnc-guidance
6. VCI. https://vci.org/about. Accessed 26 Aug 2022
7. DIVOC: DIVOC Wiki. https://divoc.egov.org.in/
8. WHO: Digital documentation of COVID-19 certificates: test result: technical specifications and implementation guidance. https://www.who.int/publications-detail-redirect/WHO-2019-nCoV-Digital_certificates_diagnostic_test_results-2022.1. Accessed 26 Aug 2022

9. PathCheck Foundation: Universal Verifier I PathCheck. https://www.pathcheck.org/en/vac cines. Accessed 26 Aug 2022

10. Zhao, Z., Ma, J.: Application of blockchain in trusted digital vaccination certificates. China CDC Wkly **4**, 106–110 (2022). https://doi.org/10.46234/ccdcw2022.021

11. Pericàs-Gornals, R., Mut-Puigserver, M., Payeras-Capellà, M.M.: Highly private blockchain-based management system for digital COVID-19 certificates. Int. J. Inf. Secur.Secur. **21**, 1069–1090 (2022). https://doi.org/10.1007/s10207-022-00598-3

12. Nabil, S.S., Alam Pran, M.S., Al Haque, A.A., Chakraborty, N.R., Chowdhury, M.J.M., Ferdous, M.S.: Blockchain-based COVID vaccination registration and monitoring. Blockchain Res. Appl. **3**, 100092 (2022). https://doi.org/10.1016/j.bcra.2022.100092

13. ESSIF-LAB: eSSIF-TRAIN by Fraunhofer-Gesellschaft I eSSIF-Lab. https://essif-lab.eu/ essif-train-by-fraunhofer-gesellschaft/. Accessed 11 Feb 2022

14. Kubach, M., Roßnagel, H.: A lightweight trust management infrastructure for self-sovereign identity. In: Schunck, R.H., Mödersheim, C.H. (eds.) Open Identity Summit 2021, Bonn Ges. Für Inform. EV, pp. 155–166 (2021)

15. Bruegger, B.P., Lipp, P.: Lightest - a lightweight infrastructure for global heterogeneous trust management (2016)

16. Jurado, V.M., Vila, X., Kubach, M., Johnson, I.H., Solana, A., Marangoni, M.: Applying assurance levels when issuing and verifying credentials using trust frameworks. In: Schunck, R.H., Mödersheim, C.H. (eds.) Open Identity Summit 2021, Bonn Ges. Für Inform. EV, pp. 167–178. 12 (2021)

17. OpenID Connect for Verifiable Presentations,. https://openid.net/specs/openid-4-verifiable-presentations-1_0.html. Accessed 11 Feb 2022

18. Jeyakumar, I.H.J., Chadwick, D.W., Kubach, M.: A novel approach to establish trust in verifiable credential issuers in self-sovereign identity ecosystems using TRAIN. Presented at the Open Identity Summit 2022, Bonn (2022)

19. TRAIN Trust List. https://s.fhg.de/gccn-registry

Privacy

Privacy-Preserving NN for IDS: A Study on the Impact of TFHE Restrictions

Ivone Amorim[1]([✉])[iD], Pedro Barbosa[2][iD], Eva Maia[2][iD], and Isabel Praça[2][iD]

[1] PORTIC – Porto Research, Technology & Innovation Center,
Polytechnic of Porto (IPP), 4200-374 Porto, Portugal
`ivone.amorim@sc.ipp.pt`
[2] Research Group on Intelligent Engineering and Computing for Advanced
Innovation and Development (GECAD), Porto School of Engineering,
Polytechnic of Porto (ISEP-IPP), 4200-072 Porto, Portugal
`{pmbba,egm,icp}@isep.ipp.pt`

Abstract. The rapid growth of the Internet ecosystem has led to an increase in malicious attacks, such as Distributed Denial of Service (DDoS), which pose a significant threat to the availability of shared services. Intrusion Detection Systems (IDSs) play a crucial role in detecting and responding to these attacks. However, the use of intelligent systems like IDSs raises significant concerns regarding privacy and the protection of network data. Homomorphic Encryption (HE) has emerged as a promising cryptographic technique for enabling privacy-preserving IDSs, but it has some limitations. In this work, we analyse the impact of HE, specifically the TFHE scheme, on the performance of Neural Networks (NNs) for DDoS attack detection, and provide the first study on assessing the impact that TFHE restrictions may have in NNs training and inference. Our findings show that TFHE restrictions have a minor impact on the performance of NNs, with the models complying with TFHE restrictions achieving performance metrics comparable to plaintext-based NNs. This suggests that high-performing NNs can be achieved, for DDoS attack detection, without exposing plaintext data. Additionally, we also observed that TFHE-compliant models exhibit a learning pace similar to traditional NNs. Therefore, our results highlight the potential of TFHE in enabling privacy-preserving NNs for DDoS attack detection, but further research is needed to gain a deeper understanding of its limitations. This may be done by exploring other metrics, and datasets, and by assessing the computational overhead in real-world scenarios.

Keywords: Privacy-preserving IDS · Distributed Denial of Service · Deep Learning · Homomorphic Encryption · TFHE · Privacy-Preserving NNs

This work was partially supported by the Norte Portugal Regional Operational Programme (NORTE 2020), under the PORTUGAL 2020 Partnership Agreement, through the European Regional Development Fund (ERDF), within project "Cybers SeC IP" (NORTE-01-0145-FEDER-000044). This work has also received funding from the project UIDB/00760/2020.

R. Rios and J. Posegga (Eds.): STM 2023, LNCS 14336, pp. 107–125, 2023.
https://doi.org/10.1007/978-3-031-47198-8_7

1   Introduction

The rapid growth of the Internet ecosystem, with an expected 500 billion devices by 2030 [14], highlights how unimaginable is our life without this interconnected system. However, all these connected devices are susceptible to various types of malicious attacks, such as Distributed Denial of Service (DDoS), which can disrupt access to shared services. Now, with Kaspersky reporting that in Q3 of 2022 there was an increase of DDoS attacks of 47.87% from Q3 of 2021 [1], it is even more urgent to create solutions capable of accurately identifying these attacks. There have been several known attacks in previous years, like in February 2021, the Crypto Currency Exchange EXMO was taken offline for 2 h following a massive DDoS attack that reached 30 gigabytes per second [25], or like in September 2022, where Activision Blizzard was a target of a DDoS attack and the outage lasted three-and-a-half hours counting from the first public acknowledgment of the attack [27]. Employing systems capable of detecting such attacks, such as Intrusion Detection Systems (IDSs), is crucial for organizations to be able to detect and respond to malicious activities effectively [34]. However, these types of intelligent systems require access to large amounts of data, some of which may be sensitive and confidential, which raises significant concerns regarding privacy and the protection of network data.

Homomorphic Encryption (HE) is a standing out cryptographic technique used in this context [9]. Its unique capability to perform computations on encrypted data without requiring decryption makes it an attractive choice. However, the adoption of HE in the context of NN for IDS also introduces significant challenges, namely because it imposes some restrictions, such as allowing only fixed-point arithmetic and having limited operations [6]. In 2021, Clet et al. [20] assessed the three most popular FHE schemes regarding their ability to allow secure NN evaluation in the cloud. The schemes considered were Brakerski-Fan-Vercauteren (BFV) [23], Cheon-Kim-Kim-Song (CKKS) [17], and fast fully homomorphic encryption scheme over the torus (TFHE) [18]. According to their findings, TFHE is the best choice when single polynomial functions are not sufficient and bitwise operations are used. On a recent literature analysis, Amorim et al. [6] also concluded that TFHE scheme achieves better results than Brakerski-Gentry-Vaikunathan (BGV) scheme [13], which is the one most used up to today. This is mainly due to its capability to represent numbers by individual bits [40], allowing for more efficient computations and expanding the range of operations that can be performed.

Despite the substantial progress that has been made, in the development of privacy-preserving NNs for IDS using HE, this research area is still in its early stages. As such, it is crucial to investigate the potential implications of HE-based NNs. Therefore, in this study, we assess the impact of TFHE restrictions on the NN's ability to learn and identify DDoS attacks. To achieve this, we follow a well-defined research methodology that includes training NN models that satisfy TFHE restrictions and assess their performance using common metrics, namely accuracy, precision, recall, and F1-score. The evaluation includes comparing the trained models with SotA NNs for DDoS detection, assessing the impact of using equal-sized but different samples from the same dataset on their performance, and analysing their learning pace across multiple epochs. Through this assessment, we obtain valuable insights into the impact that TFHE restrictions have in NN training, specifically in the domain of DDoS detection. The comparisons conducted in our work showed that the impact of TFHE restrictions is noticeable, but not that significant. Also, the use of equal-sized samples of the same dataset does not have a significant impact on the performance NNs that satisfy TFHE restrictions, that is, TFHE-compliant NNs. Finally, the learning pace of these models is very high in the first epochs, which may lead to good performance metrics and less computational overhead.

The rest of this paper is organized as follows. Section 2 introduces the main concepts in NNs, IDS, and HE. We also provide details regarding the TFHE scheme, highlighting its key characteristics. Section 3 reviews the main research work on the implications of applying HE in NNs. In Sect. 4, we present our research methodology, which includes a review of the literature on DDoS detection and the application of HE and TFHE for NN's training and inference, as further detailed in Sect. 5. The experimental tests are detailed in Sect. 6, and the analysis and discussion of our experimental results are provided in Sect. 7. Lastly, in Sect. 8, we highlight the key findings of our work and provide research directions for further investigation.

2   Background

2.1   Neural Networks

NNs are a type of Machine Learning (ML) algorithm that simulates the functionality of the human brain [28]. It is made of interconnected nodes, called *artificial neurons*, that are grouped in *layers*. These neurons output a signal which is calculated using an *Activation Function* (AF) whose input is a weighted sum of the inputs. Different AFs give different behaviour to neurons, thus the selection is made based on the specific use case. Common AFs include the Sigmoid function, usually denoted by $\sigma(x)$, ReLU (Rectified Linear Unit), and Softmax, which are defined in Table 1.

Table 1. Most common activation functions

Sigmoid	$\sigma(x) = \frac{1}{1+e^{-x}}$
ReLU	$f(x) = max(0, x) = \begin{cases} 0 & \text{if } x < 0 \\ x & \text{if } x \geq 0 \end{cases}$
Softmax	$f(x_i) = \frac{e^{x_i}}{\sum_{j=1}^{N} e^{x_j}}$

To aggregate multiple inputs, the neuron performs a weighted sum of them. To train NNs, these weights are updated in order to reduce the difference between the expected and the received output, which is done using what is called the *loss function*. Common loss functions include Mean Squared Error (MSE), cross-entropy, and binary cross-entropy.

Deep NNs (DNNs) [4] are NNs that contain more than two layers (input and output layers). They are the most general type of NN, capable of performing almost all tasks. However, in certain cases, more specialized alternatives can offer increased efficiency and better results. Convolutional Neural Networks (CNNs) are a specific type of NNs designed for analysing matrix data. They exploit the concept of convolution, where filters are applied to input data to extract features. Convolutional layers capture local patterns, and subsequent pooling layers reduce the dimensionality while preserving essential information. Recurrent Neural Networks (RNNs) are designed to process sequential data, since they possess feedback connections, allowing them to retain information from previous inputs. This recurrent architecture enables RNNs to capture dependencies and long-term context within sequential data. Long Short-Term Memory (LSTM) and Gated Recurrent Unit (GRU) are popular variants of RNNs that mitigate the vanishing gradient problem, which can hinder learning in traditional RNNs.

When evaluating the performance of NNs models, metrics are employed and each of them provides a different perspective on the model performance. Some of the more common metrics are: *accuracy, precision, recall*, and *F1-Score* [30].

2.2   Homomorphic Encryption and TFHE

HE is a type of encryption that, besides guaranteeing data privacy, allows computations to be performed directly on the ciphertext. This means that a third party can perform operations on the encrypted data without ever knowing the original plaintext. There are various categories of HE schemes, which can be classified according to the type and number of operations they enable [2]. The first HE schemes only allowed a single type of operation (known as Partial Homomorphic Encryption schemes). Consequent types started to allow multiple operations but a limited amount of times (known as Somewhat Homomorphic Encryption schemes), but only Fully Homomorphic Encryption (FHE) schemes allow unlimited computations on multiple operation types at once.

TFHE [18] is an FHE scheme that was introduced as a faster alternative to previous FHE schemes, such as BGV [12] and CKKS [17] that suffered from high computational complexity and limited functionality. The security of this scheme is based on a hard lattice problem called Learning With Errors (LWE) [36], which is a mathematical problem that forms the foundation of many lattice-based cryptographic schemes. In fact, the majority of FHE schemes used nowadays are LWE-based and use noisy ciphertexts. TFHE has several advantages, namely, it introduces several optimizations, such as a bootstrapping technique that allows this scheme to perform a wide range of computations on encrypted data, including addition, multiplication, and comparison operations. This latter operation is not possible to perform on data encrypted with other FHE schemes, such as BGV and CKKS. Consequently, TFHE has gained significant attention in both academia and industry due to its ability to perform arbitrary computations on encrypted data.

The TFHE scheme is typically used with numbers represented in the fixed-point number format. This is because the commonly used floating-point representation is too complex to be effectively employed in this scheme [40]. In their work, Song et al. [39] proposed a set of algorithms to implement arithmetic operations and comparison operations on numbers represented in the fixed-point format and encrypted using the TFHE scheme. An implementation of this scheme is already available, which includes several optimizations, and is the one used in our work[1].

3 Related Work

The use of HE to uphold data privacy in NNs has received substantial recognition within the research community, but there are few works that study the implications of applying HE in this context, namely regarding the restrictions it imposes and their impact in NN performance. In 2019, Boura et al. [10] simulated the noise resulting from homomorphic computations and function approximations, to analyse the stability of NN performance using the Homomorphic Encryption for Arithmetic of Approximate Numbers (HEAAN) [16,17] and TFHE schemes. In their method, the authors modelled the noise using Gaussian distributions. A series of experiments were conducted to evaluate the robustness of different NNs to internal weight perturbations that may arise from the propagation of significant homomorphic noise. They performed experiments on three distinct CNNs and discovered that all of them could tolerate relative errors of at least 10% without significantly impacting overall accuracy.

In 2018, Bourse et al. [11] introduced a new framework named FHE-DiNN for the homomorphic evaluation of what they called deep discretized NN. In their approach, unlike in standard NNs, the weights and biases, as well as the domain and range of the activation function, cannot be real-valued and must be discretized. This discretization process is the reason for the name "Discretized Neural Networks" or DiNNs. While their main goal was not to assess the implications of using encrypted data in NN inference, they compared the accuracy of FHE-DiNN with the accuracy of the same NN operating on plaintext data. However, they do not consider the problem of training a NN over encrypted data. Instead, they assume that the NN is trained with plaintext data and focus on the inference part. The authors also refined the TFHE scheme to improve its efficiency, albeit at the cost of increased storage requirements. In their experiments, for a network with 30 hidden neurons, they achieved a classification accuracy of 93.55% with unencrypted data and 93.71% with homomorphically encrypted data. In a NN with 100 hidden neurons, they achieved an accuracy of 96.43% with unencrypted inputs and 96.35% with encrypted inputs. The loss in accuracy was expected. However, the authors themselves suggest that to improve the

---

[1] https://www.tfhe.com/.

classification accuracy of these discretized networks, it would be interesting to train a DiNN from scratch rather than simply discretizing a pre-trained model.

Similarly to the work of Bourse et al., most of the research in this field has applied HE to the inference part. In our work, as mentioned in Sect. 1, we focus on analysing the impact of TFHE restrictions when both training and inference are performed. This approach aligns with the suggestion of Bourse et al. to train a DiNN instead of simply discretizing a pre-trained model.

4   Research Methodology

The goal of this study is to evaluate how the limitations imposed by the TFHE scheme affect NN's ability to accurately identify DDoS attacks on network data. To that purpose, we have developed a research methodology consisting of five main steps, as depicted in Fig. 1.

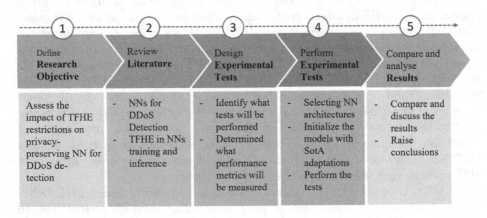

Fig. 1. Research methodology overview

The first step consists of clearly defining our *research objective*: **Assess the impact of TFHE restrictions on privacy-preserving NN for DDoS detection**. The second step comprises to *review the literature*, that includes two main topics. First, it is necessary to analyse the literature on NNs for DDoS Detection, so we can identify the SotA NN models to serve as baselines for our study, as well as the dataset to be used. Then, the literature on the use of HE to preserve data privacy on NN training and inference needs to be studied, with a focus on the use of TFHE. The goal is to identify TFHE restrictions and analyse the main approaches that have been used in the literature to address them. As a result, we can determine the required adaptations for NNs to be able to work with TFHE-encrypted data, and to identify SotA approaches to address these limitations, so that TFHE-compliant NN models can perform optimally.

Having the literature reviewed, we can move to the *design experimental tests* stage, to determine what experiments should be performed to assess the impact of TFHE restrictions on the ability of NN models to learn, when SotA adaptations are used. This step also includes defining what performance metrics will be used and what comparisons will be made. Then, we will *perform the experimental tests* identified in the previous step, by selecting suitable NN architectures, and initializing the NN models with appropriate parameters, such as the number of layers, the number of neurons in each layer and the activation functions. All of these choices will be compliant with the previously identified restrictions. Finally, we will *compare and analyse the results*, which is a fundamental step to raise conclusions regarding the impact of TFHE restrictions in NNs.

5 Literature Review

In this section, we address the second step of our research methodology. First, it is presented a review of DDoS detection using NNs and the most used datasets. Then, it is reviewed the application of HE to NNs training and inference, focusing on TFHE scheme, its restrictions, and the best approaches to address the consequent limitations.

5.1 NNs for DDoS Detection and Datasets

DDoS detection mechanisms are extremely important due to the increasing prevalence of this kind of attacks and their potential impact. A lot of work has already been devoted to address this problem. Ali et al. [3] conducted a comprehensive systematic review that analyses the application of ML techniques in detecting DDoS attacks within Software-Defined Networking (SDN) environments. Additionally, Mittal et al. [30] presented a study more focused on Deep Learning (DL) approaches for DDoS attack detection. NNs stand out by playing a significant role, with 29.41% of the works analysed by Mittal et al. utilizing CNNs and 29.59% leveraging DNNs. In fact, approaches using NNs are the most common in recent years [3].

The work conducted by Sbai and El Boukhari [37] proposes an IDS for DDoS attacks specifically targeting data flooding, utilizing DNNs and the CICD-DoS2019 dataset [38]. They reported a binary classification accuracy of 99.997% for UDP-based attacks. CNN models have also been explored to detect complex DoS and DDoS attacks by converting network traffic datasets into images. This approach was employed by Hussain et al. [26]. They proposed a methodology to convert network traffic data into image format and trained a SotA CNN model, specifically ResNet, using the converted data. According to their claims, their approach achieved a remarkable 99.99% accuracy in binary classification for detecting DoS and DDoS attacks. The dataset used was again CICDDoS2019.

In 2021, Assis et al. [7] proposed a SDN defense system that relies on the analysis of individual IP flow records. The system utilizes a GRU DL method for detecting both DDoS and intrusion attacks. The authors compared the model results with other ML approaches, including DNN, CNN, k-Nearest Neighbors (kNN), using CICDDoS2019 and CICIDS 2018 datasets. The results conducted on the CICDDoS2019 dataset demonstrated that most of the tested methods achieved comparable performance, with the kNN and GRU approaches achieving the highest accuracy rates for legitimate flow classification, at 99.7% and 99.6%, respectively. On the other hand, using the CICIDS 2018 dataset, the results of the methods varied more significantly, and the GRU approach achieved an accuracy of 97.1%. Also, in 2021, Cil et al. [19] utilized a DNN to detect DDoS attacks in a sample of packets captured from network traffic. Their experiments were conducted on the CICDDoS2019 dataset. They claimed that their model successfully detected DDoS attacks in network traffic with a 99.99% accuracy, and accurately classified the attack types with an accuracy of 94.57%. Amaizu et al. [5] proposed a novel two-DNN approach, one for identifying the attacks, and the other for classifying the attack types. They achieved an accuracy of 99.66% using the CICDDoS2019 dataset.

It is notable that most of the works published after 2020 have achieved a reported accuracy of over 99 %. Additionally, the majority of these works utilized the CICDDoS2019 dataset. This dataset is the most up-to-date and contains a larger number of samples compared to other network traffic datasets, such as KDD-99, NSL-KDD, DEFCON, CAIDA, CICIDS2017, and UNSW-NB15. Moreover, the CICIDS2017 and CICDDoS2019 datasets are the preferred choices in nearly half of the literature on DDoS attack detection [30]. Despite the known errors in CICFlowMeter tool, which may lead to some flaws in the widely used CICDDoS2019 dataset, to the best of our knowledge, no corrections have been provided to CICDDoS2019 dataset, unlike the CICIDS2017 dataset [22]. As such, in this work we use the CICDDoS2019 dataset.

5.2 HE and TFHE in NNs Training and Inference

Several scientific works have proposed the use of HE to preserve data privacy in NNs. One of the most relevant is the work by Dowlin et al. [21], who introduced the concept of CryptoNet, which is a modified version of a trained NN to operate on encrypted data. However, this work does not use encrypted data in the training process. Other approaches have been proposed in the literature [35], but, according to Amorim et al. [6], few have explored the use of HE for NN training and classification. Their work also highlighted that the main restrictions on the application of HE to NNs are: limited operations, computational overhead, and data representation. The former is related to the limitation of HE schemes regarding the operations they allow to be performed over encrypted data, and their consequences on non-linear AFs like the sigmoid, and pooling techniques which require comparing values. The computational overhead limitation is related to the high cost of performing operations over encrypted data. Finally, HE schemes do not represent the data all in the same way. Typically, they operate on encrypted data in a fixed-point representation, which is different from the floating-point representation used in traditional NNs. This may negatively affect the accuracy and performance metrics. Amorim et al. also concluded that BGV scheme is the most commonly used in this context. However, TFHE arose as a surprising approach to reduce training times while maintaining high accuracy rates, because of its ability to allow defining more complex operations in a bitwise manner, including comparison.

The first work to use the TFHE scheme to create a trainable NN was conducted by Lou et al. and it was named "Glyph" [29]. Their approach uses BGV and TFHE schemes together to accelerate HE operations. TFHE was employed to enable the use of non-linear activation functions like ReLU and Softmax, with the latter being done with table lookup operations. The authors performed several experiments and concluded that their approach achieved a similar accuracy to SotA model [32], but in much less time. However, it is worth noting that the authors did not formally compare the accuracy of Glyph with SotA NN models operating on plaintext data.

The second work that uses TFHE is the work by Yoo et al. [40], and, contrary to the previous one, this only uses the TFHE scheme. This framework was named "t-BMPNet". To represent the numbers, they used fixed-point number representation, and they implemented the operations based on the work by Song et al. [39]. To better approximate the sigmoid activation function, they proposed an algorithm to calculate the exponential function in the cyphertext space. The authors claim that this framework allows for training in the encrypted domain without needing to perform polynomial approximations. They conducted multiple experiments to validate their approach and found that t-BMPNet successfully achieved a highly accurate design of the non-linear sigmoid function compared to other methods that rely on polynomial approximations. It is worth noting that they did not compare the performance metrics of t-BMPNet with plaintext-based NNs. Also, the authors' implementation of the exponential function demonstrates weaknesses that limit its practical application. Therefore, in our work, we conduct a comprehensive literature review focusing on the approximation of the sigmoid AF and the SotA approach addressing this limitation.

In the work of Dowlin et al. [21], the sigmoid AF was replaced with the square function, which does not resemble the typical sigmoid shape. Other approaches adopted higher degree polynomials as activation functions for more training stability. For instance, Onoufriou et al. [33] replaced the sigmoid AF with a polynomial approximation with degree 3, namely

$$\sigma(x) \approx 0.5 + 0.197x - 0.004x^3,$$

which, according to the authors, closely follows the standard sigmoid between the ranges of -5 and 5. Yuan et al. [41] approximated the sigmoid activation function using MacLaurin series expansion:

$$\sigma(x) = \frac{1}{2} + \frac{x}{4} + \frac{x^2}{48} + \frac{x^5}{480} + O(x^6).$$

Other approaches include the work of Ghimes et al. [24] which used the simplest activation function possible, the identity function, to reduce the cost of computing over encrypted data. A remarkable approach is the use of piecewise approximations, which can be computed with data encrypted using TFHE but is not supported by the other most common FHE schemes, since they do not allow comparison operations [18]. One such example is the piecewise approximation by Myers et al. [31] which is used in the work of Zhang et al. [15] and in the work of Bansal et al. [8]. This piecewise approximation combines 9 linear functions, that follow the sigmoid behaviour in the interval from −8 to 8.

$$
\begin{cases}
1 & x > 0 \\
0.015625x + 0.875 & 4 < x \leq 8 \\
0.03125x + 0.8125 & 2 < x \leq 4 \\
0.125x + 0.625 & 1 < x \leq 2 \\
0.25x + 0.5 & -1 < x \leq 1 \\
0.125x + 0.375 & -2 < x \leq -1 \\
0.03125x + 0.1875 & -4 < x \leq -2 \\
0.015625x + 0.125 & -8 < x \leq -4 \\
0 & x \leq -8
\end{cases}
\tag{1}
$$

According to Zhang et al., this piece-wise approximation is the best-performing approximation in NNs [42]. Furthermore, the fact that this approximation only requires the calculation of simple linear functions makes it less expensive to compute than polynomial approximations. As such, in our work, this was the selected approximation to address the non-linearity of sigmoid AF.

6   Experimental Tests

6.1   Design

Following the reviewed literature presented in the previous section, it was possible to identify two SotA NN models for the detection of DDoS attacks. The works of Cil et al. [19], and Sbai and El Boukharei [37] are the ones with the best performance to detect DDoS attacks, which is recognized by the reviews of Ali et al. [3] and Mittal et al. [30]. As such, these two models will serve as the baselines for our study. Let N_C be the NN model suggested by Cil et al., and N_S the model of Sabi and El Boukharei.

To perform our analysis, two NN models will be trained in compliance with TFHE restrictions and using the mentioned SotA approaches to address those restrictions. We will perform three different types of tests, and our analysis will be done using the most common performance metrics used in this context, namely accuracy, precision, recall and F1-score [3,30].

Test 1: The first test involves comparing the performance of models N_C and N_S with the performance of our trained NN models that incorporate TFHE adaptations. This will allow us to assess if the required adaptations enable us to achieve a similar level of performance, or if there is a significant decrease. For this test, we will utilize a CICDDoS2019 sample that is of the same size as N_C and N_S, but distinct from them. Consequently, we will also analyse the impact of sampling on the performance of the trained models.

Test 2: This test involves training several NN models, both with and without TFHE adaptations, using equal-sized but different samples of the CICD-DoS2019 dataset. The training will be conducted for 5 epochs, which will provide sufficient insights into its potential impact.

Test 3: The third test aims to compare the learning pace of our best trained NN model with and without the TFHE-required adaptations. The models will be trained for 15 epochs, and performance metrics will be computed at the end of each epoch. This analysis will help us determine if TFHE adaptations influence the learning pace of the models.

6.2 Execution

We have trained two TFHE-compliant NN networks to compare with N_C and N_S, respectively. Let N'_C be our NN which is compared with N_C, and N'_S the one which is compared with N_S. First, we will describe the architecture of our networks: N'_C is composed of 5 layers, with the input layer being composed of 69 units, one for each feature. Each hidden layer is made of 50 neurons, and the final layer is composed of 1 neuron. The only difference between this architecture and the one of N_C is the number of neurons of the output layer, which is 2 in N_C; N'_S is composed of 3 layers. The input and hidden layers have 7 neurons, and the final layer is composed of 1 neuron because it is a binary classification. The input size considered was 7 (4 continuous features plus 3 protocol types because we used the one-hot-encoding strategy – a technique applied to transform categorical inputs into a numerical form which represents each category as a separate feature).

Regarding the initialization and training of these NNs, it was done considering the test to be performed:

Test 1: The number representation was defined to be fixed-point for N'_C and N'_S. The activation function of all layers was defined to be the sigmoid approximation presented in (1). Notice that the TFHE scheme, as previously mentioned, requires the use of fixed-point number representation, since floating-point representation is not appropriate [40]. However, fixed-point representation is less flexible because it does not automatically adjust to the precision of the represented number, which is the number of bits used to represent the fractional part of the number. In fixed-point represented numbers, this number of bits allocated must be manually specified, resulting in a less dynamic representation.

To address this issue, firstly, we normalize the data and choose an appropriate number of bits to represent the fractional part. Additionally, because of the bitwise nature of the TFHE scheme, manual implementation of arithmetic operations was necessary to achieve optimal performance. After reviewing the SotA literature, we have chosen to implement TFHE operations according to the definition provided by Song et al. [39]. At the end of this process, all the required operations are ready to be used in the TFHE-compliant NNs N'_C and N'_S.

Test 2 and 3: Let N_C'' be the NN with the same architecture as N_C', but such that its number representation is floating-point, and the AF is the sigmoid. This NN is trained in the TensorFlow library[2] with the usual operations, and their performance metrics are compared with the ones of N_C'.

The data used to train and test N_C' and N_C'' is a sample of the original CICD-DoS2019 dataset, in a similar process to the one performed by Cil et al. [19]. Each sample is made up of exactly 365474 entries, were 80% is for training and 20% is for testing. To increase the performance of the final model, the features "Flow ID", "SourceIP", "SourcePort", "DestinationIP", "DestinationPort", "Protocol", "Timestamp", and "SimilarHTTP" were discarded due to their lack of contribution to the final model [19]. The features "Bwd PSH Flags", "Fwd URG Flags", "Bwd URG Flags", "Fwd Bytes/Bulk Avg", "Fwd Packet/Bulk Avg", "Fwd Bulk Rate Avg", "Bwd Bytes/Bulk Avg", "Bwd Packet/Bulk Avg", and "Bwd Bulk Rate Avg" were also removed since they all contained a single value. All the 69 remaining features are numeric, and their normalization was done using the Min-Max normalization. As previously said, the classification is binary with the labels being either Benign or Malign.

The data used in N_S' and N_S'' followed a similar normalization process. But, in this case, the dataset entries chosen were the ones referring to benign traffic and to the UDP attacks. Only 4 features from the original dataset were used, namely "Destination Port", "Packet Length Std", "Packet Length Std", "min seg size forward", and "Protocol", because they were defined as the most impacting for UDP attacks by Sharafaldin et al. [38]. Since one of the selected features is categorical (the "Protocol" feature), there was the need to apply the one-hot-encoding technique to transform it to a numerical feature, which resulted in a total of 7 features.

7   Compare and Analyse the Results

7.1   TFHE-Compliant Models vs. SotA NNs for DDoS Detection

The comparison between our trained models with SotA models N_C and N_S is crucial to understand the extent to which TFHE restrictions, and our mitigation measures, affect their performance to correctly classify DDoS attacks. A sample of the CICDDoS2019 dataset was generated as detailed in the previous section, and a total of 15 epochs were used to train our models. Table 2 presents the performance metrics obtained and the corresponding Relative Error (RE).

Examining the results of N_C' when compared with N_C, it can be observed that there was a minor decrease across all metrics. The largest decrease is observed in accuracy, with an RE of 0.1%, followed by recall with an RE of 0.09%. Similarly, when comparing the values obtained for N_S and N_S', we can observe that the errors are higher than the ones obtained with the other NN. In this case, accuracy and recall are again the metrics with a higher RE, 0.96% and 2.03%, respectively.

---

[2] https://www.tensorflow.org/.

Table 2. Performance metrics and relative error.

Metric	N_C	N_C'	RE	N_S	N_S'	RE
Accuracy	99.97%	99.87%	0.10%	99.995%	99.04%	0.96%
Precision	99.99%	99.96%	0.03%	99%	98.74%	0.26%
Recall	99.98%	99.89%	0.09%	98%	99.99%	2.03%
F1-Score	99.98%	99.92%	0.06%	99%	99.36%	0.36%

Regarding the recall's RE, it is worth mentioning that it does not represent a loss, since our model obtained a better performance (99.99%) than N_S (98%).

In summary, all the changes introduced for our NN to comply with TFHE restrictions cause a maximum RE difference of 0.09% from the work of Cil et al. [19]. On the other hand, the maximum relative difference between N_S' and the DNN by Sbai and El Boukhari [37] has a maximum difference of 2%. However, this difference has to be interpreted with caution since, in the metrics recall and F1-score, N_S' performs better than N_S. These results allow us to conclude that, at least in DDoS attack detection, the adaptations required to make DNNs compatible with TFHE have a negligible effect on the NN's ability to learn and generalize the data provided. From this, we may conclude that we can achieve high performing NN models for DDoS detection without exposing the plaintext data to third parties since they can work directly with encrypted data.

7.2 Impact of Equal-Sized But Different Samples on NN Model's Performance

The experiments conducted in the SotA NN models for DDoS attack detection did not use the entire CICDDoS2019 dataset, as explained before. Therefore, we trained our NN models using equal-sized but different samples from the same dataset. We then compared the performance metrics of N_C' and N_C'' at the end of five epochs. The purpose of this experiment was to assess if using different samples of the dataset would result in significant variations in the obtained metrics. Figure 2 shows the performance metrics obtained for each NN model and each sample of the CICDDoS2019 dataset.

From the results obtained, it is possible to observe that the sample variation has a higher impact in the TFHE-compliant model, N_S', than in N_C''. However, none of the metrics exhibit significant variations, which can be validated by the low values of the standard deviation obtained for each metric, as presented in Table 3. Therefore, we may conclude that the sample variation does not have a significant impact on the performance of the model obtained from N_S', meaning that the restrictions imposed by TFHE, when SotA approaches are used to address those limitations, are not significant. The results obtained in this experiment, and presented in Fig. 2, also allow comparing the performance of N_C' with the performance of N_C''. It is possible to observe that overall, the performance metrics of N_C' are worse than the corresponding metrics in N_C'', which is aligned

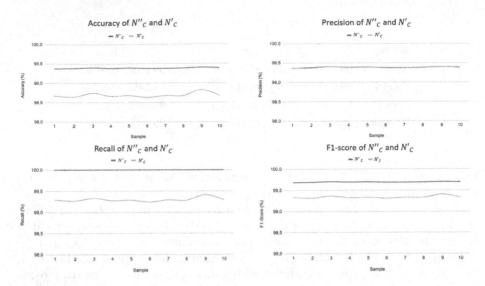

Fig. 2. Performance metrics obtained in N'_C and N''_C for ten different samples of CICD-DoS2019

with the comparison done in the previous test, where N'_C showed a lower performance when compared with N_C. It can also be observed that the metric with the smallest difference between the means is precision, which is not surprising considering the chart corresponding to this metric in Fig. 2. This indicates that N'_C was able to correctly classify attack entries in almost the same proportion as N''_C. The metric recall also shows a low difference between the means, although that difference is larger than the difference in precision. Both accuracy and F1-score metrics have a very similar difference in means, around 0.7%, which can be significant in some application scenarios. On the good side, with just 5 epochs, the TFHE-compliant NN model was able to achieve high-performance metrics. Specifically, the accuracy consistently exceeds 98.6%, precision exceeds 99.3%, F1-score exceeds 99.2%, and recall exceeds 99.3%.

In conclusion, while the choice of sample had a higher impact on the TFHE-compliant model, there were no significant variations in the performance metrics of either model. Also, while N'_C got slightly worse performance metrics compared to N''_C, it still achieved high accuracy, precision, F1-score, and recall. These results highlight the potential of TFHE in privacy-preserving NNs.

7.3   Learning Pace Analysis

The learning pace of a model is another important factor that should be considered, especially when our objectives include training NN models with TFHE-encrypted data. In these models, the cost of performing operations is much higher than performing the usual operations over plaintext data. Consequently, it is important to assess how fast TFHE-compliant models achieve good

Table 3. Performance metrics mean and standard deviation for ten different samples and five epochs

Metric	N'_C		N''_C		Mean
	Mean	S. deviation	Mean	S. deviation	Difference
Accuracy	0.986907449	0.000561599	0.993835419	0.00010242	0.00692797
Precision	0.993819436	0.000118783	0.993823498	0.000125635	0.4062×10^{-5}
Recall	0.993001315	0.00047724	1.00	0.00	0.006998685
F1-score	0.993410171	0.000284419	0.996899313	6.32853×10^{-5}	0.003489142

performance metrics and compare their results with plaintext-based NN models. A high learning pace means that few epochs will allow achieving high-performance metrics, suggesting that fewer operations are required. To analyse and compare the learning pace of N'_C and N''_C, each of these NNs was trained using the same sample of CICDDoS2019 and underwent 15 epochs to evaluate the learning pace. The metric accuracy was computed at the end of each epoch, and Fig. 3 summarizes the results. From this figure, it is evident that N'_C has a higher learning pace in the first epoch, but then it decreases, and the performance slowly increases until epoch 15. On the other hand, N''_C demonstrates a rapid increase between epochs 2 and 3, followed by a similar behaviour to that of N'_C. Again, in this experiment, it is clear that TFHE-compliant NN model demonstrates inferior performance when it comes to accuracy. Also, something which is important to highlight is the fact that the learning pace of N'_C is much higher in the beginning, which suggests that few epochs may be sufficient to achieve a good performance.

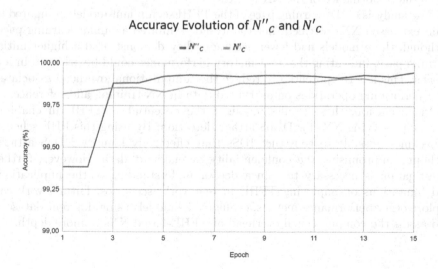

Fig. 3. Performance metrics obtained in N'_C and N''_C for 15 different epochs

In conclusion, TFHE-compliant models show a similar learning pace when compared with plaintext-based models. Moreover, our experiments suggest that despite having lower accuracy, N'_C may achieve satisfactory performance in few epochs. This is beneficial as it helps reduce the number of operations required to be performed on encrypted data and, consequently, the total computational overhead.

8   Conclusions and Future Work

In this work, we conducted an analysis of the impact of TFHE restrictions on the ability of NNs to learn and accurately identify DDoS attacks. For that, we performed a comparison between TFHE-compliant models and SotA NNs, which revealed that the adaptations required to make the NNs compatible with TFHE had a minor effect on their ability to learn and generalize the data provided. In our best-performing model, the maximum relative error difference observed was 0.1% for the metric accuracy, which is not considered a significant difference in this context. Furthermore, this model also demonstrated comparable performance metrics, regarding precision, recall, and F1-score, which indicates that high-performing NN models for DDoS detection can be achieved without exposing plaintext data to third parties.

Since different samples of the same dataset were used by the different works, we also investigated the impact of using different samples of CICDDoS2019 dataset on the performance of TFHE-compliant models. Our results suggest that the sample did not result in significant variations in the TFHE-compliant models' performance, nor in plaintext-based ones. This indicates that the restrictions imposed by TFHE, when mitigated using SotA approaches, do not significantly affect the performance of the NN models.

The analysis of the learning pace of the TFHE-compliant models compared to plaintext-based NNs revealed that the former exhibited a similar learning pace. Although these models had lower accuracy, they demonstrated a higher initial learning pace, indicating that satisfactory performance could be achieved in few epochs. This has the potential to reduce the computational overhead associated with performing operations on encrypted data in NN training and inference.

In conclusion, these results highlight the potential of TFHE in enabling privacy-preserving NNs for DDoS attack detection. By using this FHE scheme, it becomes possible to construct IDSs that effectively identify DDoS attacks without compromising the confidentiality of plaintext data. However, further investigation is necessary to gain a deeper understanding of the implications and limitations of employing TFHE in real-world scenarios. Future work can explore other performance metrics, evaluate the models using different datasets, and assess the computational overhead of TFHE-based NNs in more depth.

References

1. Hacktivists step back giving way to professionals: a look at DDos in Q3 2022 (2022). https://www.kaspersky.com/about/press-releases/2022_hacktivists-step-back-giving-way-to-professionals-a-look-at-ddos-in-q3-2022. Accessed 30 Jun 2023

2. Acar, A., Aksu, H., Uluagac, A., Conti, M.: A survey on homomorphic encryption schemes: Theory and implementation. ACM Comput. Surv. **51**(4), 1–35 (2018)

3. Ali, T.E., Chong, Y.W., Manickam, S.: Machine learning techniques to detect a DDos attack in SDN: a systematic review. Appl. Sci. **13**(5), 3183 (2023). https://www.mdpi.com/2076-3417/13/5/3183

4. Alom, M.Z., et al.: A state-of-the-art survey on deep learning theory and architectures. Electronics **8**(3), 292 (2019)

5. Amaizu, G., Nwakanma, C., Bhardwaj, S., Lee, J., Kim, D.: Composite and efficient DDos attack detection framework for b5g networks. Comput. Netw. **188**, 107871 (2021). https://doi.org/10.1016/j.comnet.2021.107871

6. Amorim, I., Maia, E., Barbosa, P., Praça, I.: Data privacy with homomorphic encryption in neural networks training and inference (2023). https://doi.org/10.48550/arXiv.2305.02225

7. Assis, M.V., Carvalho, L.F., Lloret, J., Proença, M.L.: A GRU deep learning system against attacks in software defined networks. J. Netw. Comput. Appl. **177**, 102942 (2021). https://doi.org/10.1016/j.jnca.2020.102942

8. Bansal, A., Chen, T., Zhong, S.: Privacy preserving back-propagation neural network learning over arbitrarily partitioned data. Neural Comput. Appl. **20**, 143–150 (2011). https://doi.org/10.1007/s00521-010-0346-z

9. Boulemtafes, A., Derhab, A., Challal, Y.: A review of privacy-preserving techniques for deep learning. Neurocomputing **384**, 21–45 (2020). https://doi.org/10.1016/j.neucom.2019.11.041

10. Boura, C., Gama, N., Georgieva, M., Jetchev, D.: Simulating homomorphic evaluation of deep learning predictions. In: Dolev, S., Hendler, D., Lodha, S., Yung, M. (eds.) CSCML 2019. LNCS, vol. 11527, pp. 212–230. Springer, Cham (2019). https://doi.org/10.1007/978-3-030-20951-3_20

11. Bourse, F., Minelli, M., Minihold, M., Paillier, P.: Fast homomorphic evaluation of deep discretized neural networks. In: Shacham, H., Boldyreva, A. (eds.) CRYPTO 2018. LNCS, vol. 10993, pp. 483–512. Springer, Cham (2018). https://doi.org/10.1007/978-3-319-96878-0_17

12. Brakerski, Z., Gentry, C., Vaikuntanathan, V.: Fully homomorphic encryption without bootstrapping. Cryptology ePrint Archive, Paper 2011/277 (2011). https://eprint.iacr.org/2011/277

13. Brakerski, Z., Gentry, C., Vaikuntanathan, V.: (Leveled) fully homomorphic encryption without bootstrapping. In: Proceedings of the 3rd Innovations in Theoretical Computer Science Conference, pp. 309–325. ITCS 2012, Association for Computing Machinery, New York, NY, USA (2012)

14. Carey, E., Donnell, I.M.: Powering an inclusive, digital future for all (2023). https://newsroom.cisco.com/c/r/newsroom/en/us/a/y2023/m01/powering-an-inclusive-digital-future-for-all.html. Accessed 29 Jun 2023

15. Chen, T., Zhong, S.: Privacy-preserving backpropagation neural network learning. IEEE Trans. Neural Networks **20**(10), 1554–1564 (2009). https://doi.org/10.1109/TNN.2009.2026902

16. Cheon, J.H., Han, K., Kim, A., Kim, M., Song, Y.: Bootstrapping for approximate homomorphic encryption. In: Nielsen, J.B., Rijmen, V. (eds.) EUROCRYPT 2018. LNCS, vol. 10820, pp. 360–384. Springer, Cham (2018). https://doi.org/10.1007/978-3-319-78381-9_14
17. Cheon, J.H., Kim, A., Kim, M., Song, Y.: Homomorphic encryption for arithmetic of approximate numbers. In: Takagi, T., Peyrin, T. (eds.) ASIACRYPT 2017. LNCS, vol. 10624, pp. 409–437. Springer, Cham (2017). https://doi.org/10.1007/978-3-319-70694-8_15
18. Chillotti, I., Gama, N., Georgieva, M., Izabachène, M.: TFHE: fast fully homomorphic encryption over the torus. J. Cryptol. **33**(1), 34–91 (2020)
19. Cil, A.E., Yildiz, K., Buldu, A.: Detection of DDos attacks with feed forward based deep neural network model. Expert Syst. Appl. **169**, 114520 (2021). https://doi.org/10.1016/j.eswa.2020.114520
20. Clet, P.-E., Stan, O., Zuber, M.: BFV, CKKS, TFHE: which one is the best for a secure neural network evaluation in the cloud? In: Zhou, J., et al. (eds.) ACNS 2021. LNCS, vol. 12809, pp. 279–300. Springer, Cham (2021). https://doi.org/10.1007/978-3-030-81645-2_16
21. Dowlin, N., Gilad-Bachrach, R., Laine, K., Lauter, K., Naehrig, M., Wernsing, J.: CryptoNets: applying neural networks to encrypted data with high throughput and accuracy. In: Proceedings of the 33rd International Conference on International Conference on Machine Learning, vol. 48. pp. 201–210. ICML 2016, JMLR.org (2016)
22. Engelen, G., Rimmer, V., Joosen, W.: Troubleshooting an intrusion detection dataset: the CICIDS2017 case study. In: 2021 IEEE Security and Privacy Workshops (SPW), pp. 7–12 (2021). https://doi.org/10.1109/SPW53761.2021.00009
23. Fan, J., Vercauteren, F.: Somewhat practical fully homomorphic encryption. Cryptology ePrint Archive, Paper 2012/144 (2012). https://eprint.iacr.org/2012/144. Accessed 24 Feb 2023
24. Ghimes, A.M., Vladuta, V.A., Patriciu, V.V., Ioniţă, A.: Applying neural network approach to homomorphic encrypted data. In: 2018 10th International Conference on Electronics, Computers and Artificial Intelligence (ECAI), pp. 1–6 (2018)
25. Haworth, J.: UK cryptocurrency exchange EXMO knocked offline by 'massive' DDos attack. https://portswigger.net/daily-swig/uk-cryptocurrency-exchange-exmo-knocked-offline-by-massive-ddos-attack (2021). Accessed 29 Jun 2023
26. Hussain, F., Abbas, S.G., Husnain, M., Fayyaz, U.U., Shahzad, F., Shah, G.A.: IoT dos and DDos attack detection using ResNet. In: 2020 IEEE 23rd International Multitopic Conference (INMIC), pp. 1–6 (2020)
27. James, M.: The 8 most devastating DDos attacks of 2022 and what we can learn from them. https://allaboutcookies.org/the-worst-ddos-attacks. Accessed 29 Jun 2023
28. Lippmann, R.P.: An introduction to computing with neural nets. SIGARCH Comput. Archit. News **16**(1), 7–25 (1988)
29. Lou, Q., Feng, B., Fox, G.C., Jiang, L.: Glyph: fast and accurately training deep neural networks on encrypted data. Neural Inf. Process. Syst. Found. **33**, 9193–9202 (2020)
30. Mittal, M., Kumar, K., Behal, S.: Deep learning approaches for detecting DDos attacks: a systematic review. Soft. Comput. (2022). https://doi.org/10.1007/s00500-021-06608-1
31. Myers, D., Hutchinson, R.: Efficient implementation of piecewise linear activation function for digital VLSI neural networks. Electron. Lett. **25**, 1662–1663 (1989). https://digital-library.theiet.org/content/journals/10.1049/el_19891114

32. Nandakumar, K., et al.: Towards deep neural network training on encrypted data. vol. 2019-June, pp. 40–48. IEEE Computer Society (2019)
33. Onoufriou, G., Mayfield, P., Leontidis, G.: Fully homomorphically encrypted deep learning as a service. Mach. Learn. Knowl. Extr. **3**, 819–834 (2021)
34. Ozkan-Okay, M., Samet, R., Aslan, d., Gupta, D.: A comprehensive systematic literature review on intrusion detection systems. IEEE Access **9**, 157727–157760 (2021). https://doi.org/10.1109/ACCESS.2021.3129336
35. Pulido-Gaytan, B., et al.: Privacy-preserving neural networks with homomorphic encryption: challenges and opportunities. Peer-to-Peer Netw. Appl. **14**, 1666–1691 (2021)
36. Regev, O.: On lattices, learning with errors, random linear codes, and cryptography. J. ACM **56**(6), 84–93 (2009)
37. Sbai, O., El Boukhari, M.: Data flooding intrusion detection system for MANETs using deep learning approach (2020). https://doi.org/10.1145/3419604.3419777
38. Sharafaldin, I., Lashkari, A.H., Hakak, S., Ghorbani, A.A.: Developing realistic distributed denial of service (DDos) attack dataset and taxonomy. In: 2019 International Carnahan Conference on Security Technology (ICCST), pp. 1–8 (2019). https://doi.org/10.1109/CCST.2019.8888419
39. Song, B., Yoo, J., Hong, M., Yoon, J.: A bitwise design and implementation for privacy-preserving data mining: from atomic operations to advanced algorithms. Secur. Commun. Netw. **2019**, 3648671 (2019). https://doi.org/10.1155/2019/3648671
40. Yoo, J.S., Yoon, J.W.: t-BMPNet: trainable bitwise multilayer perceptron neural network over fully homomorphic encryption scheme. Secur. Commun. Netw. **2021**, 7621260 (2021). https://doi.org/10.1155/2021/7621260
41. Yuan, J., Yu, S.: Privacy preserving back-propagation neural network learning made practical with cloud computing. IEEE Trans. Parallel Distrib. Syst. **25**(1), 212–221 (2014). https://doi.org/10.1109/TPDS.2013.18
42. Zhang, Q., Xin, C., Wu, H.: SecureTrain: an approximation-free and computationally efficient framework for privacy-preserved neural network training. IEEE Trans. Netw. Sci. Eng. **9**(1), 187–202 (2022). https://doi.org/10.1109/TNSE.2020.3040704

Analyzing and Improving Eligibility Verifiability of the Proposed Belgian Remote Voting System

Jan Willemson[(✉)] [ID]

Cybernetica, Narva mnt 20, 51009 Tartu, Estonia
jan.willemson@cyber.ee

Abstract. This paper discusses a recent hybrid paper-electronic voting system proposal put forward for Belgian elections. We point to some problems in the proposal, and consider addition of blind signatures as one possible approach to dealing with the identified shortcomings. We discuss the concomitant updates from both the protocol and electoral processes point of view, arguing that our proposal would strike a better balance between different requirements. To the best of our knowledge, this is also the first proposal to use blind signatures in a paper-based voting system.

Keywords: Remote voting · eligibility verification · blind signatures

1 Introduction

In our current increasingly mobile world, it becomes harder and harder to get all the eligible voters to physical polling stations for the act of voting on a single day [27]. The recent COVID-19 outburst has only added to this problem [9]. Hence, there is a definite need for reliable remote voting options.

Two main approaches are available for this. The more established way is to send the ballots in via physical mail. For example in the 2020 U.S. presidential elections, 43% of the voters cast their ballot by mail – a number twice as high as four years earlier. Even though the COVID pandemic was definitely a major factor, the trend towards increasing voting by mail has been observed for years in the U.S. [10].

As an alternative, several countries like Switzerland [14], Estonia [12], Norway [25], Australia [8], France [6], etc. have had elections with vote casting options over Internet.

Both of these approaches have their pros and cons. Internet voting can offer reliable vote transmission and efficient tallying procedures. On the other hand, it has been criticized for implementation complexity, concentrating many risks into the central components, being hard to verify by an average citizen, etc. [15,21,24]

Postal voting can be implemented without relying on any digital equipment on the client side, hence being easy to understand, use and trust by the voter.

R. Rios and J. Posegga (Eds.): STM 2023, LNCS 14336, pp. 126–135, 2023.
https://doi.org/10.1007/978-3-031-47198-8_8

On the other hand, it is very hard to ensure authenticity and privacy of the voters, the postal channel is vulnerable to both integrity and confidentiality attacks, etc. [4,17]

Thus, it is natural to ask whether we could get the best of the both worlds without sacrificing too much in terms of residual risks. And indeed, several digital-paper hybrid schemes have been proposed in the literature [3,4,10,20]. Of course, building such a hybrid system requires trade-offs, and balancing different requirements may lead to several possible equilibria.

In this paper, we are going to take a closer look at a recent proposal presented by a team of Belgian researchers with the aim of being implemented for postal voting in Belgium [1,2,11]. The advantage of this proposal over the previous purely academic papers is that it comes with technical implementation details much better laid out. Belgium also has a national electronic identity system which opens up new opportunities in terms of voter authentication and eligibility verification.

We note that the current version of the paper has been shortened due to space limitations, but interested readers can have access to the full version at [28]. The main difference between the current and the full version is a review and analysis of previously proposed hybrid schemes, more specifically the Benaloh-Ryan-Teague scheme [4], STROBE [3], RemoteVote and SAFE Vote by Crimmins et al. [10], and the scheme by McMurtry et al. [20]. Remarkably, none of the proposals explicitly deals with eligibility verification, even though it is one of the core components of end-to-end verifiability.

Crimmins et al. claim in [10] that STROBE, RemoteVote and SAFE Vote all provide the eligibility verification property, but do not specify how exactly. The only explanation given is a reference to 'existing procedural controls' in a footnote, possibly hinting at the standard methods used in postal voting like double envelopes.

Note, however, that double envelope system is a legacy adopted not because of its excellent properties, but because historically there has not been a better alternative. For instance it does not really protect vote secrecy against a malicious actor while the vote is in transit; thus we question the ballot secrecy claims made in Table 1 of [10]. This is a good example that one can not leave any part of the system unspecified while proposing a new voting scheme as implementation details of one component may harm the desired properties of others.

2   The Proposed Belgian Remote Voting System

The system proposed for Belgium relies on verification codes that have to be recorded by the voter in order to perform the verification later [1,2,11]. More precisely, the voter is provided with three sheets (see Fig. 1). The selection sheet lists all the candidates together with the preference marking spots. The code sheet presents short codes for both of the options of voting for or against a particular candidate. Finally, the note sheet is meant for the voter to write down the codes corresponding to her selections in order to later check against the codes published on the bulletin board. As the code sheet provides a receipt of voting, it is meant to be destroyed after the vote has been cast.

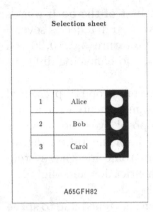
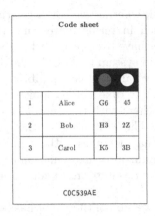
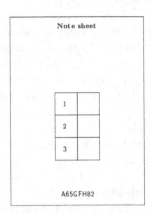

Fig. 1. Ballot, code and note sheet for the proposed Belgian postal voting system [1, 2, 11]

What makes the case of the proposed Belgian system especially interesting is the existence of a well-established national electronic identity (eID) infrastructure. It means that voter authentication can be performed much more reliably, potentially also improving eligibility verification in the case of postal voting.

In the beginning of the process, the voter logs onto the voting server by using her eID. The server checks eligibility and prepares the voting sheets specifically for this voter. Each sheet will carry a random 128-bit voter-specific code k, enabling the tallying authority to authenticate the vote even without relying on an outer, signed envelope. This allows, in principle, to drop the outer envelope altogether, thus potentially increasing postal vote secrecy during the vote transmission.

In Belgium, the selection sheets can be pretty large as they need to accommodate all the candidates. However, the voter can only vote for the candidates of one party. Thus, as a compromise, in case of electronically prepared ballots, it is proposed that the candidates of one party are displayed on one A4 paper, and the voter would need to mail in only the sheet corresponding to the party of her choice.

It has been left unspecified in the system description [1] whether the voter can only print out the sheet she needs for her party of choice, or whether she should print out all the generated sheets. The subtle issue here is keeping the vote secret from the voter's computer. If the voter would choose to print the candidate list of only one party, a malicious device would learn her party preference.

There are a few ways to look at the issue. On one hand, voter's device definitely is one of the easiest-to-attack components in the whole system, especially when it comes to vote secrecy. Even if the system provides vote integrity verification mechanisms, it is hard to give strong guarantees that the vote has not been leaked from the used digital device. The best known mechanism to achieve such guarantees would be code voting, but it comes with usability trade-offs [19] and we do not consider such systems in this paper.

A usual approach to this problem is not to target vote secrecy at all in remote settings, and aim at a weaker property of coercion resistance instead (see e.g. [18] for an overview of different proposed approaches to achieve it). Since the Belgian system has been presented as the first step of transition towards Internet voting, there will be a moment in the future when the voters will use their computing devices to prepare and cast votes. Thus we argue that leaking one's vote to the computer is a practical trade-off that will need to be accepted at some point anyway.

We note that in the Belgian system as it is described in [1], the voter has to trust her computer also in regards to vote integrity. It is foreseen that the voter can contact the ballot preparation server and check that the code k is a valid one, but it is not guaranteed to be unique. If an attacker is able to compromise several voter devices, he can make these devices to use the same (valid!) k for all the ballots issued through them. This problem would only be noticed in the tallying phase and the system description [1] does not specify what to do in this case. However, there are little alternatives to invalidating all the votes sharing the same k, as the tallying authority can not distinguish the k-sharing-attack from a ballot box stuffing attempt. This efficiently results in disenfranchising all the voters who cast these votes.

We may try to detect multiple verification attempts made to the same code k, but it is unclear what to do in case of successful detection. The voter may legitimately want to verify the code several times from different devices as she does not necessarily trust a single device. Also, most of the voters would probably not bother verifying the code at all, and thus such a detection mechanism would likely be inefficient.

We also note that checking the value of k for validity may pose a usability issue. The system description [1] discusses embedding k on the ballot sheets both in an OCR font and in the form of a QR code, recommending the former to support human readability. In both cases, the voter would need a device capable of scanning the representation of k, which in the current practice means having a smartphone, a tablet computer or alike. In any case the success of scanning depends on the user skills, quality of the camera, lighting conditions, etc.

3 Eligibility Assurance with Blind Signatures

The root of the problem enabling reuse of the values for k is that these values depend neither on the voter, nor the vote. Of course we do not want to print the voter's digital signature on the ballot instead as this would undermine vote secrecy. Luckily, there exists a good alternative available in the form of blind signatures.

Blind signatures were first introduced by Chaum in 1982 in the context of implementing untraceable payment systems [7]. In 1992, Fujioka et al. proposed using this primitive to achieve vote-secrecy-preserving authentication of a ballot by blindly signing it with the authority's key [13].

The construction of Fujioka *et al.* is a very generic one, with a number of improvements proposed throughout the years (see e.g. [23] for a good overview on the topic). Blind signatures have been used also in practical e-voting schemes; recently e.g. in Russia [26]. However, their practical applicability to postal voting has been very limited. This can be explained by a diverse set of assumptions that one would need in order for make such a solution to be useful and work.

On one hand, in order to make use of blind signatures, a relatively advanced digital infrastructure is required. The voters need a reliable means for authenticating themselves to the signing authority, accompanied with a method to do something with the returned signature. On the other hand, even though a digital identification infrastructure is assumed, the society should still look to improve remote paper vote casting, rather than going for Internet voting right away.

Both of these aspects are present in Belgium, and hence considering the blind signatures for authenticity and eligibility assurance is interesting in this case.

Of course, we would need to use the voter's computer as a ballot marking device and trust it for vote secrecy. However, as discussed in Sect. 2, this is a trade-off that is probably required sooner or later anyway.

Thus, we propose setting up a generic blind signature scheme as an addition to the proposed Belgian postal voting system. For that, we will assume the authority A who maintains the list of eligible voters, possesses a public-private blind signature key pair, and publishes the corresponding public key.

After the voter has used her computer to fill in the ballot, it is first masked for blind signing. The voter then authenticates herself to A who verifies her eligibility. If this verification succeeds, A issues the blind signature. Next, the voter's computer removes the blinding and displays the obtained signature directly on the ballot, e.g. as a QR code. The resulting sheet can then be printed out and cast as a regular postal ballot.

Before mailing it off, this scheme allows the voter to check well-formedness of the ballot and the signature of A. First of all, note that the Belgian ballot can be encoded rather efficiently. There are less than 256 parties running, so one byte is enough to encode the party choice. For each candidate of this party, one bit needs to be encoded. Depending on the length of the the party list, one may need a few dozens of bits. Adding the metadata concerning the election event, the encoding should comfortably fit into 256 bits.

This means that we can put the padded encoding of the vote directly under the signature without hashing it. Thus, a mobile verification app can be developed that can decode the whole vote together with A's signature from the QR code, check the signature and display the decoded vote content to the voter. The voter can then visually match the result to what has been printed out on the ballot in the traditional human-readable way. This ensures the voter that the vote has indeed been correctly signed by A without intermediate manipulation.

Machine-readable votes also allow for a more efficient tallying process by scanning the QR codes. It is not even necessary to visually inspect all the postal ballots for correspondence to the human readable part if a proper statistical post-election audit process like risk-limiting audit is implemented. Note that a statistical post-election audit as part of the tallying procedures implicitly also protects the voters who did not bother downloading and using the mobile verification app.

As the ballot is signed with the authority's signature to prove eligibility, there is no need for the outer, voter-identifying envelope, and the ballot can be mailed anonymously. This removes one of the major privacy problems of postal voting that anyone can study the envelopes in transit and reveal how the postal voters voted. At the same time, blind signatures printed on the ballots ensure eligibility of the voters and also protect ballot integrity.

On the other hand, extra measures are then needed at the polling station on the election day. If a person who has cast a postal vote comes to the polling station and wants to cast a vote, a respective mechanism is needed to avoid double voting.

In case of a standard double envelope postal voting system (see e.g. [16]), envelopes can be kept sealed until the regular polling station votes are also cast. Double envelopes belonging to the voters who submitted in-person votes can then be discarded without opening.

In case of anonymously sent postal votes this approach would not work. Instead, the voter needs to be stopped at the polling station before she gets a chance to submit a vote. For that, polling stations workers need access to A's database of voters who have requested signing their postal votes. This is technically non-trivial, but doable. A similar system has been in use in Estonia since 2021 Parliamentary elections to enable cancelling electronic votes with paper ones in a polling station [12].

Note that the problem of double voting is also present in the proposed Belgian system as described in [1], and even on a bit more serious level. In principle, the ballot preparation server can keep a list of the voters who have requested a ballot, but it can not tell if the ballot has actually been completed. If requesting a blank ballot would be registered as the voter having used her voting rights, this may end in disenfranchising the voter e.g. in the case she fails submitting her postal vote and attempts voting in a polling station.

In case of our proposal, the voter only requests the authority's signature *after* having filled the ballot in. Of course, we still do not know whether the signed ballot was actually mailed or not. However, we argue that there is a potential legal difference between just requesting a blank ballot and asking for the authority's confirmation once it has been filled. In the latter case it is easier to call the act of voting completed and rule against the voter in case of a possible dispute between disenfranchisement *vs.* double voting.

Using the voter's computer as a ballot marking device also allows for a more efficient printing procedure. There is no reason to print all the sheets corresponding to the parties the voter did not want to vote for. Of course, this is mainly a result of our trade-off with secrecy of the vote from the voter's computer. On the other hand, it also gives a significant environmental effect as the number of otherwise unused sheets of paper would be multiplied by the number of postal voters.

Note also that we do not need to print the code sheet at all. Instead, we can directly generate and print the filled note sheet. This is good both from the usability and security points of view. Usability benefits are clear as the voter is not required to copy any random codes by hand. Security benefit comes from the observation that the code sheet is actually a receipt that the voter can use intentionally or under coercion to prove how she voted.

The original system description [1] requires the voter to destroy this sheet, but we argue that relying on such a measure to achieve privacy properties is not a good security design principle. Users can in general only be expected to give a minimal amount of effort to achieve the functional goals, i.e. casting one's vote in our case. If the code sheet remains lying around, it can cause unexpected privacy problems which are better avoided if possible.

4   Discussion

Verifiability properties of standard double envelope postal voting are rather weak. There is typically no Cast as Intended verification, and instead of Counted as Cast there is a weaker property of Counted as Collected [5]. We argue that if such a system would be proposed today, it would not be accepted as not satisfying elementary requirements, especially as postal voting protocols offering better properties are available now [1, 3, 4, 10, 20, 29].

However, eligibility verification remains a challenge for all these proposals, and this problem is inherently related to the available infrastructure. When we want to enable e.g. Cast as Intended verification, we need to enhance the capabilities of the verifier, i.e. the voter. When postal voting was introduced for the soldiers fighting in the U.S. Civil War, there was no way of getting convenient and fast feedback about the fate of the vote [22]. But nowadays we have omnipresent Internet access, enabling such feedback.

A similar situation also occurs for eligibility checking. However, now the primary verification agent is the election organizer who needs to decide whether the vote came from a legitimate voter, and whether it is a double vote or not. The ballot can carry some sort of an identifier (like a social security number), or the outer envelope may carry a signature, but neither of them can be considered a strong form of identification in the third decade of the 21st century.

In order to provide better eligibility verification properties, a respective infrastructure is required. With electronic identity mechanisms being introduced in many countries, this infrastructure is becoming readily available. It is only natural to use it to secure remote voting, both in electronic and mail-in settings.

The most straightforward way of integrating an eID into a remote voting scheme would be signing the vote. In case of electronic voting it is easy to encrypt the vote in order to protect its confidentiality. For postal voting, however, there is an implicit expectation that the paper representation of the vote should be human-readable. This makes direct signing with voter's eID impossible.

On the other hand, blind signature on an anonymous paper vote is still very much an option. Of course, a corrupt signing authority may attempt to sign the votes for ineligible voters. As a solution, we can require the blind signing requests to be signed by the voters. If in the end of the voting period the number of authority-signed votes in the digital ballot box exceeds the number of voter-signed requests then we know that the authority has cheated. As an alternative, signing authority can be implemented in a distributed manner in order to avoid relying on just one trusted party.

We also note that while the idea of using blind signatures in a remote voting setting is not novel, their application to paper-based voting systems is to the best of our knowledge.

5   Conclusions

Cryptographically-enhanced postal voting is a recent and exciting research area. It has a potential to provide a remote voting solution with better authentication and integrity properties compared to regular postal voting. At the same time, it can also avoid some of the problems with remote electronic vote casting as the main vote carrying medium would still be paper.

In this article we reviewed several recent schemes, concentrating on the system proposed by Belgian researchers as an intermediate step towards Internet voting. It adds end-to-end verification capabilities to the postal votes and can even be used to send the filled ballots in anonymously.

The proposal is very rich in implementation details compared to previous purely academic papers. It is also very realistic in terms of the trade-offs required between usability, verifiability and privacy properties of the system.

However, we were still able to point out several problems in this paper. The biggest issue is the need to trust the voter's computer not to disenfranchise the voter by maliciously re-using the random authentication token k.

In order to mitigate this problem, we proposed implementing a generic blind signature scheme instead of using the random token. It turns out that such a solution also has other benefits; for example it enables easier tallying and vote correctness verification by the voter.

The downside of our proposal is the need to use the voter's PC as a ballot preparation device, hence trusting the device not to breach vote secrecy. However, we argue that this is a reasonable trade-off that will need to be accepted at some point anyway. At the same time we reduce the need for paper print-outs. This improves both the environmental footprint and coercion-resistance properties of the scheme.

The Belgian postal voting scheme is still in the early stages of research, and we hope that this paper has made a small contribution towards its future success.

Acknowledgments. The paper has been supported by the Estonian Research Council under the grant number PRG920.

References

1. Étude sur la possibilité d'introduire le vote Internet en Belgique (2021). https:// elections.fgov.be/informations-generales/etude-sur-la-possibilite-dintroduire-le-vote-internet-en-belgique
2. Abeels, T.: Postal Voting. Master's thesis, Ecole polytechnique de Louvain, Université catholique de Louvain (2021). http://hdl.handle.net/2078.1/thesis:33143
3. Benaloh, J.: STROBE-Voting: send two, receive one ballot encoding. In: Krimmer, R., et al. (eds.) E-Vote-ID 2021. LNCS, vol. 12900, pp. 33–46. Springer, Cham (2021). https://doi.org/10.1007/978-3-030-86942-7_3
4. Benaloh, J., Ryan, P.Y.A., Teague, V.: Verifiable postal voting. In: Christianson, B., Malcolm, J., Stajano, F., Anderson, J., Bonneau, J. (eds.) Security Protocols 2013. LNCS, vol. 8263, pp. 54–65. Springer, Heidelberg (2013). https://doi.org/10.1007/978-3-642-41717-7_8
5. Bernhard, M., et al.: Public evidence from secret ballots. In: Krimmer, R., Volkamer, M., Braun Binder, N., Kersting, N., Pereira, O., Schürmann, C. (eds.) E-Vote-ID 2017. LNCS, vol. 10615, pp. 84–109. Springer, Cham (2017). https://doi.org/10.1007/978-3-319-68687-5_6
6. Blanchard, E., Gallais, A., Leblond, E., Sidhoum-Rahal, D., Walter, J.: An analysis of the security and privacy issues of the Neovote online voting system. In: Krimmer, R., Volkamer, M., Duenas-Cid, D., Rønne, P., Germann, M. (eds.) Electronic Voting. E-Vote-ID 2022. LNCS, vol. 13553. Springer, Cham (2022). https://doi.org/10.1007/978-3-031-15911-4_1
7. Chaum, D.: Blind Signatures for Untraceable Payments. In: Chaum, D., Rivest, R.L., Sherman, A.T. (eds.) CRYPTO 1982, pp. 199–203. Plenum Press, New York (1982). https://doi.org/10.1007/978-1-4757-0602-4_18
8. Conway, A., Teague, V.: iVote Issues: assessment of potential impacts on the 2021 NSW local government elections. In: Proceedings of E-Vote-ID 2022, pp. 42–52 (2022)
9. Cotti, C., Engelhardt, B., Foster, J., Nesson, E., Niekamp, P.: The relationship between in-person voting and COVID-19: evidence from the Wisconsin primary. Contemp. Econ. Policy **39**(4), 760–777 (2021)
10. Crimmins, B.L., Rhea, M., Halderman, J.A.: RemoteVote and SAFE vote: towards usable end-to-end verification for vote-by-mail. In: Matsuo, S., et al. Financial Cryptography and Data Security. FC 2022 International Workshops. FC 2022. LNCS, vol. 13412. Springer, Cham (2023). https://doi.org/10.1007/978-3-031-32415-4_27
11. Devillez, H.: Secure postal voting. In: Proceedings of E-Vote-ID 2022, pp. 140–143 (2022). https://dspace.ut.ee/handle/10062/84432
12. Ehin, P., Solvak, M., Willemson, J., Vinkel, P.: Internet voting in Estonia 2005–2019: evidence from eleven elections. Gov. Inf. Q. **39**(4), 101718 (2022). https://doi.org/10.1016/j.giq.2022.101718
13. Fujioka, A., Okamoto, T., Ohta, K.: A practical secret voting scheme for large scale elections. In: Seberry, J., Zheng, Y. (eds.) AUSCRYPT 1992. LNCS, vol. 718, pp. 244–251. Springer, Heidelberg (1993). https://doi.org/10.1007/3-540-57220-1_66
14. Haines, T., Pereira, O., Teague, V.: Running the race: a swiss voting story. In: E-Vote-ID 2022, Proceedings. LNCS, vol. 13553, pp. 53–69. Springer (2022). https://doi.org/10.1007/978-3-031-15911-4_4
15. Halderman, J.A.: Practical attacks on real-world e-voting. In: Real-World Electronic Voting, pp. 159–186. Auerbach Publications (2016)

16. Killer, C., Stiller, B.: The swiss postal voting process and its system and security analysis. In: Krimmer, R., et al. (eds.) E-Vote-ID 2019. LNCS, vol. 11759, pp. 134–149. Springer, Cham (2019). https://doi.org/10.1007/978-3-030-30625-0_9

17. Krimmer, R., Volkamer, M.: Bits or Paper? Comparing Remote Electronic Voting to Postal Voting. In: EGOV 2005. Schriftenreihe Informatik, vol. 13, pp. 225–232. Universitätsverlag Rudolf Trauner, Linz, Austria (2005)

18. Krips, K., Willemson, J.: On practical aspects of coercion-resistant remote voting systems. In: Krimmer, R., et al. (eds.) E-Vote-ID 2019. LNCS, vol. 11759, pp. 216–232. Springer, Cham (2019). https://doi.org/10.1007/978-3-030-30625-0_14

19. Marky, K., Schmitz, M., Lange, F., Mühlhäuser, M.: Usability of code voting modalities. In: Conference on Human Factors in Computing Systems, CHI 2019. ACM (2019). https://doi.org/10.1145/3290607.3312971

20. McMurtry, E., Boyen, X., Culnane, C., Gjøsteen, K., Haines, T., Teague, V.: Towards Verifiable Remote Voting with Paper Assurance (2021). https://doi.org/10.48550/ARXIV.2111.04210

21. Park, S., Specter, M., Narula, N., Rivest, R.L.: Going from bad to worse: from internet voting to blockchain voting. J. Cybersecurity 7(1), 1–15 (2021)

22. Rotondi, J.P.: Vote-by-mail programs date back to the civil war (2020). https://www.history.com/news/vote-by-mail-soldiers-war

23. Schmid, M., Grünert, A.: Blind Signatures and Blind Signature E-Voting Protocols (2008), University of Applied Science Biel, Switzerland. https://www.e-voting-cc.ch/images/pdf/blindsignatures.pdf

24. Simons, B.: Why internet voting is dangerous. Geo. L. Tech. Rev. 4, 543–563 (2019)

25. Stenerud, I.S.G., Bull, C.: When reality comes knocking Norwegian experiences with verifiable electronic voting. In: EVOTE 2012. LNI, vol. P-205, pp. 21–33. GI (2012). https://dl.gi.de/20.500.12116/18219

26. Vakarjuk, J., Snetkov, N., Willemson, J.: Russian Federal Remote E-voting Scheme of 2021 - Protocol Description and Analysis. In: EICC 2022, pp. 29–35. ACM (2022). https://doi.org/10.1145/3528580.3528586

27. Willemson, J.: Bits or paper: which should get to carry your vote? J. Inf. Secur. Appl. 38, 124–131 (2018). https://doi.org/10.1016/j.jisa.2017.11.007

28. Willemson, J.: Analyzing and Improving Eligibility Verifiability of the Proposed Belgian Remote Voting System. CoRR abs/2305.09411 (2023). https://doi.org/10.48550/arXiv.2305.09411

29. Zagórski, F., Carback, R.T., Chaum, D., Clark, J., Essex, A., Vora, P.L.: Remotegrity: design and use of an end-to-end verifiable remote voting system. In: Jacobson, M., Locasto, M., Mohassel, P., Safavi-Naini, R. (eds.) ACNS 2013. LNCS, vol. 7954, pp. 441–457. Springer, Heidelberg (2013). https://doi.org/10.1007/978-3-642-38980-1_28

Consent as Mechanism to Preserve Information Privacy: Its Origin, Evolution, and Current Relevance

Marietjie Botes[✉] [iD]

Stellenbosch University, Cape Town, South Africa
wmbotes@sun.ac.za

Abstract. Informed consent and the requirements to obtain ethical-legal sound consent has a long and rich history that originated with the medical treatment of patients and then evolved into its application in the field of biomedical research. The same concepts and principles of consent has been adopted to be applied in the digital sphere. However, upon closer scrutiny it is clear why this principle, that originated for the protection of a person's bodily integrity cannot be adequately applied in the digital sphere to protect people's personal data. To the contrary it transpired that the ethical-legal requirements of consent has been made futile in the context of digital consent receipts by erroneously comparing and applying this concept to transactions receipts and commercial contracts. This paper investigates this evolution of biomedical consent to digital consent and analyze the difference between the concept of consent as it developed for biomedical application and compare that with the current application of consent in the digital sphere.

Keywords: Consent · consent receipt · information privacy · personal data control · information autonomy

1 Introducing Informed Consent

The first appearance of the requirement of informed consent in a legally binding document was between 1891–1900 in Germany. In 1891, the Prussian Minister of Interior Affairs issued a directive to all prisons, in which he specified that tuberculosis treatment must not be administered against the will of the medical patient. Another ministerial directive was issued in 1900 to all hospitals and clinics in the country which excluded all minors or otherwise non-competent persons from non-therapeutic clinical studies and required the provision of "unambiguous consent" after the proper disclosure of the negative consequences of the study to all participants [1].

The informed consent standard was first internationally recognized in the Nuremberg Code in response to the controversial and criminal research activities conducted on concentration camp prisoners at the end of World War II. The Nuremberg tribunal defined the essential nature of voluntary consent from a human participant in any type of medical experiment, delegitimizing research with humans who are incapable of providing consent, and stressed that "any element of force, fraud, deceit, duress, overreaching,"

R. Rios and J. Posegga (Eds.): STM 2023, LNCS 14336, pp. 136–145, 2023.
https://doi.org/10.1007/978-3-031-47198-8_9

"constraint" or "coercion" should not be part of valid consent [2]. Also, the participant "should have sufficient knowledge and comprehension about the elements" of the study (nature, duration, purpose, method, and meaning of the experiment) involved [2]. This includes the knowledge of any inconveniences, hazards, and effects, which should be expected. The duty and responsibility for ascertaining that the consent is valid rests upon the "individual who initiates, directs, or engages in the experiment" and it cannot be delegated to anybody else [2]. Later, this code obstructed the research and development of novel therapies for serious medical conditions [3] and was further developed by the WMA's Helsinki Declaration [4], which distinguished non-therapeutic from therapeutic research, permitting under special circumstances the latter also on incompetent persons [5].

Beauchamp and Childress [6] divide informed consent into three consent threshold elements: 1) competence such as voluntariness; 2) information elements such as disclosure, provision of recommendations, and understanding; and 3) consent elements such as decision-making and authorization. The provision of adequate in-formation about the objectives, risks, and benefits of any intervention is pivotal to the ethical practice of medicine. However, these information requirements were soon disregarded by doctors who conducted longitudinal Syphilis studies on patients in Tuskegee from 1932–1972 without their knowledge or consent [7]. Consequently to improve informed consent practices and to better protect an individual's ability to exercise control and decisional power over their bodies, the Belmont Report was issued as an outcry against the paternalistic practices that dominated doctor–patient relationships with the goal of enabling individuals to exercise control and decisional power over their bodies [8].

2   Evolution into Digital Consent

The Menlo Report, which is globally regarded as the ethical framework for research involving Information and Communications Technologies, explicitly states that it is based on the principles of the Belmont Report, which resulted from the Tuskegee Syphilis research scandal to specifically expand and refine principles around biomedical informed consent [9]. The Menlo Report bases the consent process on three elements: 1) information, 2) comprehension; and 3) voluntariness. However, a simple checking of these elements does not equate to ethical digital consent, neither does it provides guidance for how much and what sort of information must be provided. It is thus prudent and logical to use the evolution of biomedical consent to guide the implementation of digital consent. A major difference between biomedical and digital consent is that where biomedical consent focused on research participant protection, digital consent focuses on information control.

Digital environments pose significantly different challenges to the concept and execution of the consent process as mechanism used to preserve information privacy than the physical biomedical environments discussed above. To offer a usable consent mechanism on the internet, an "open notice and consent receipt architecture" has been proposed by Lizar and Hodder to serve as a public data management control tool for digital consent [10]. They believe that the development of a standard consent receipt schema linked to legally required consent notices may "open up control of personal data in a simple but usable way" [10]. Their Open Notice Initiative, launched in 2013, premises their consent receipt concept on a market-based approach [11]. According to this approach con-sent receipts, as in the case with regular money-based transaction receipts, can be easily aggregated in a digital format and will allow individuals to 'autonomously manage consent preferences and make choices on aggregate in the same way as companies provide transaction receipts to their customers [10]. However, such a receipt does not facilitate the exchange of any information, as required in ethical consent, or provides any indication of the terms and conditions of the commercial agreement that led to the agreed-upon transaction between the parties [13].

A comparison and critical analysis of this concept of consent receipts and the legal-ethical use of consent as a mechanism to pre-serve digital information privacy against the historical-ethical evolution of consent, as discussed above, poses fundamental misconceptions about the essence of ethical consent and what the purpose of so-called receipts in transactions is. Ethical consent re-quires the provision of enough information in an understandable way to enable proper consideration of options to lead to fully in-formed consent [12]. A receipt in commercial transactions simply proves that a transaction between two parties has occurred, for example, that money was paid in exchange.

2.1 Information Exchange

The Open Notice Initiative acknowledges that 'information autonomy' is one of the key components to enable personal data control on the internet [10]. The practical result of this was the creation of the Consent Receipt Specification by the Kantara Initiative which constitutes an interoperable record that presents metadata and context associated with the consent given by a person, which record is available to both the data subject (user) and data controller (web manager) [14]. In essence, this mechanism towards personal data control constitutes a technically interoperable record, whereas autonomy in the context of informational privacy requires the exchange of information between the parties to enable decision-making on the side of the user. Although this consent receipt may technically facilitate the online management of consent records, in its current form it still presents significant gaps. The consent receipt does not provide for any authentication or verification of the identities of the parties involved, or from a legal-ethical perspective, the exchange of any information between the parties [15]. Proactive participation may be considered a technical gap, but data controllers cannot shy away from their legal obligations in this regard for the purpose of obtaining consent, because data controllers are legally obligated to provide the prescribed information 'in a concise, transparent, intelligible and easily accessible form, using clear and plain language' [16]. But, despite these identified shortcomings, the larger argument for using digital consent receipts is to serve as proof and provide a record of successfully obtained consents to ultimately establish trust through transparency and accountability [15].

Supplementary to the legal-ethical requirements for consent, the ISO/IEC 29184:2020 standard for Online Privacy Notices and Consent provides more technically orientated specifications to 'shape the content and the structure of online privacy notices as well as the process of asking for consent to collect and process personally identifiable information (PII)' [17]. In this regard, it is important to note that a document that specifies technical standards is not legally binding in nature and cannot overrule ethical or legal obligations for the provision of specific information to users to enable legal-ethical consent. Accordingly, the controls mentioned in this ISO document may well shape the 'structure of online privacy notices' or the online 'process of asking consent', but it cannot 'shape the content' of such privacy notices or the consenting process, as the content is prescribed as mentioned before. Because this ISO document reflects a lot of the already legally prescribed requirements applicable when requesting consent (such as the provision of clear, concise, transparent, intelligible, easily accessible and understandable information in plain language about personal data processing), for purposes of regulatory harmonization it would have been better if technical standards like the ISO could simply refer to already existing legal requirements and use widely accepted and used terminology such as 'personal data' as defined in the GDPR, instead of 'personally identifiable information (PII)' as per this ISO document. This is particularly important considering that this ISO standard has been published in June 2020, being a date after the enactment of the GDPR in 2018, which provided definitions for the basic terminology used in the context of data privacy and consent. To prevent duplication and misunderstandings of different definitions attributed to similar concepts such as personal data, it is advisable that future standards use existing definitions and provisions already contained in legislation, and simply expand on the technical aspects and standards necessary to implement the regulatory requirements practically and effectively. This need for a 'common privacy terminology' was already identified in the ISO/IEC 29100 Lead Privacy Framework [18]. The primary aim of this standards framework is to provide guidance to organizations on how to protect the personal information of users for the safeguarding of privacy within an Information and Communication Technology system (ICT).

A 'common privacy terminology' is also critically important when deciding the type of consent to obtain to comply with privacy laws such as the GDPR. Personal information that includes information about one's racial or ethnic origin, political opinions, religious or philosophical beliefs, trade-union membership, the processing of genetic data, biometric data for the purpose uniquely identifying a natural person, data concerning health, or data concerning a natural person's sex life or sexual orientation qualify as 'special' personal information which requires explicit consent in terms of article 9 of the GDPR. In these circumstances explicit consent can only be obtained through a statement that must 'specify the nature of da-ta that's being collected, the details of the automated decision and its effects, or the details of the data to be transferred and the risks of the transfer' [19]. In other words, explicit consent requires a presentation of an explicit statement regarding the specific person-al data to be collected and requires an explicit action to be taken by the user to provide lawful consent, such as physically ticking a box that states 'I consent' before such consent will be compliant. In contrast, neither the Menlo Report, nor the ISO/IEC 29100 Lead Privacy Framework provide for 'special' personal information. The ISO/IEC 29100 Lead Privacy Framework only provides guidance regarding 'personally identifiable in-formation (PII)' and the exercising of explicit or express consent 'through an affirmative act indicating such consent' by the user but is silent on standards with regards to consent for 'special' personal information. Subsequently, to ensure that technical standards also contribute to the safeguarding of privacy via consent models, it is critical that ICT research regulations and online privacy consent standards ensure that they also provide for 'special' personal information as contemplated in the GDPR.

However, even if consent is only requested for the processing of non-special personal information (which does not require explicit consent), very few users are aware of, or comprehend the 'policies' and 'notices' shown online to whom they are requested to give their consent to, or details about the third parties with whom their personal information will be shared with, or what such parties will really do with their personal information [20]. This façade of available privacy notices on the internet which fails to provide adequate information has been called out as the 'biggest lie on the internet' [21]. This phenomenon of 'dark patterns' is manipulating and coercing consent from users on large scale without the option or ability to scrutinize any information about third party recipients of their information [22]. To protect users from potential online exploitation via the use of their personal data, the recently enacted California Consumer Protection Act (CCPA) now provides for a mechanism that allow users to 'opt-out' of the 'selling' of their online data to third parties and mandates the provision of a 'do-not-sell' option on all websites [23].

2.2 Comprehension

The Belmont report, on which the Menlo Report is based, states that researchers are obligated to ensure that participants understand or comprehend information provided to them [24]. It is thus insufficient and unethical to only provide information without making sure that participants understand the information. If participants cannot under-stand information, they are unable to adequately consider such information, and will the decision they base thereon be half-informed, or worse, a guess in the dark. Although the Belmont and Menlo reports specifically pertain to research, the requirement to ensure the comprehension of information by people to allow them to inform their online decisions is also mandated in the GDPR [25] and equally applicable to digital consent processes – hence the large body of research on dark patterns and manipulative technologies being used to interfere with users' online decision-making abilities [26]. In this context, it is important to note that a user's maturity, capacity for understanding, language, literacy, and more specifically his or her digital literacy must be considered when information is being presented with the aim of obtaining informed consent [27].

Subsequently, the way in which information is being presented to users is equally important to obtaining legal-ethical consent, as the information conveyed to the user. In a study by Nouwens et al. in which they investigated how some of the most com-mon consent management platform designs affect users' consent choices, they found that although the notification style (banner or barrier) had no effect, the removal of the opt-out but-ton from the first page increased consent by 22–23 percentage points, and by provid-ing more granular controls on the first page decreased consent by 8–20 percentage points [28]. Users' ability to provide legal-ethical con-sent is accordingly also affected by these technical processes, which directly link interface designs to legal-ethical compliance. In Planet49 v Bundesverband der Verbraucherzentralen und Verbraucherverbände the Court of Justice ruled that these types of designs were not a valid form of consent, not in terms of the GDPR or previous laws dating from the mid-90s [29]. Similarly, the UK's Information Commissioner's Office said that any consent mechanism that highlights the term 'agree' or 'allow' over 'reject' or 'block' constitutes a non-compliant approach to obtaining consent, because the web manager is actively trying to influence users to click on the 'accept' option [30].

But not all people are blessed with the same abilities. Some people's ability to comprehend certain types of information or information in general may be severely limited due to age, conditions of immaturity, educational level, language barriers, disability, or other factors. In these circumstances, special care must be taken to determine whether such a person will be able to sufficiently understand the information given to them to ultimately allow for an informed decision or consent. In the context of biomedical research consent, it seems easier to determine individual or group vulnerability, based, for example, on belonging to an ethnic minority group [31], or being an orphaned child [32], which groups are more prone to being exploited or harmed. In these cases, researchers are obligated to take extra precautions to protect these participants by adjusting information to a level and format that the participant will find accommodating, and most importantly comprehendible [33].

However, online vulnerability is much harder to determine. Online commerce and the collection of users' online information exposes them to the use and possible misuse of their personal data by organizations unknown to them and often without obtaining their consent for such purposes [34]. In addition to interface designs and other dark patterns that are constantly trying to manipulate users into consenting, online vulnerability is also fueled by users' inability to confirm the identity of third parties using their information and for what purpose, which poses a direct risk to their privacy and, in the context of online commerce, their financial wellbeing, which in turn critically influences their free choice to give consent. In this regard, Kaspersky.com reported that 2% of global online transactions in 2019 were fraudulent, whilst 16% of transactions were categorized as high risk because they entailed access from unauthorized parties [35]. In this context vulnerability thus manifests as a lack of control due to the restriction to the user's consumer experience and satisfaction [36], and the imbalance in the relationship between the user and the web manager because of the power which the web manager holds over the user by also controlling the knowledge, information, and quality of service that the user receives [37].

Lastly, vulnerability in this context is also a changing phenomenon that is sensitive and reactive to dynamic changes in market conditions and users' personal circumstances [38]. Subsequently, any extra protection required by regulatory instruments when dealing with vulnerable people or groups must also take a much broader concept of vulnerability into account [39], which ironically places a larger burden of providing clear and understandable information to users on the shoulders of web managers, as opposed to that of biomedical researchers, because biomedical vulnerability circumstances (as mentioned above) is not as flexible as that of online users. However, identifying some users as vulnerable seems meaningless in the absence of any strategies to enable vulnerable consumers to obtain additional skills and mitigate their risks [40]. Furthermore, not the Menlo Report, GDPR, or the recently adopted European Data Protection Board guidelines on Dark Patterns in Social Media Platform Interfaces contain any provisions that specifically deal with online vulnerability and how information should be presented to vulnerable users for purposes of obtaining consent, save for briefly referring to children as a vulnerable population who needs additional safeguards [41]. This legal lacuna needs urgent investigation. In this context, consent receipts can at best be understood as supportive technical tools for the management of consent records.

2.3 Voluntariness

Autonomy as an important legal-ethical element of consent is based on the freedom to decide free from coercion or undue influence. In a biomedical research context coercion has been described as occurring when an 'overt threat of harm is intentionally presented by one person to another in order to obtain compliance' and undue influence as occurring 'through an offer of an excessive, unwarranted, inappropriate or improper reward or other overture in order to obtain compliance' [24]. In contrast, coercion or undue influence in the digital context are often exercised by so-called persuasive computing technologies, manifesting for example in the use of dark patterns as mentioned above. The effects of these technologies are defined by the pioneer of persuasive technologies, BJ Fogg, as 'computing systems, devices, or applications intentionally designed to change a person's attitudes or behavior in a predetermined way' [42]. Although the intention to persuade users must be present, the actual detection of these technologies is often absent. Susser, Roesler, and Nissembaum describe the essence of online manipulative practices' goal to influence or persuade, but without being detected [43]. Manipulation of an individual's ability to exercise an informed decision lawfully and ethically is thus hidden, or it entails the covert subversion of a person's decision-making power by targeting or exploiting a user's decision-making vulnerabilities [44]. Manipulation greatly impacts a user's ability to either con-sider or comprehend information in these circumstances, because it keeps vital information from users which deprives them of being able to properly consider their options to provide lawful and ethical consent. It completely disrespects the autonomy of the decision-maker and may ultimately change how and what users decide leading to possible changes in their online behavior [45].

3 Current Relevance and Conclusion

The aim of all the above-discussed consent mechanisms is to give control back to individuals over their own bodies (via biomedical consent initially) and their own personal data. But by stipulating the legal obligations of the web manager in legislation such as the GDPR may certainly motivate lawful consent, but not necessarily ethical consent and true individual control and autonomy as intended. However, by simply ticking all the legal requirement boxes, and generating a consent receipt to track consent giving, will be considered legal, but does not help the user to take control of his in-formation, decision-making processes, and ultimately his life as these consent mechanisms and regulations advocate for. The GDPR is silent about any obligations of the web manager or data controller to ensure that users fully comprehend the information provided to them, contrary to the ethical requirements in the case of biomedical research, which requirements have evolved into digital consent requirements as adopted by the Menlo Report.

Digitization of consent often leads to over- or hyper-automation which may undermine the control that users have as contemplated in the GDPR. Digital consent tools must aim to empower online users to better understand consent or privacy notifications and what platforms intend to do with any personal information they collect. Although consent management tools are a vibrant area of research, research projects focusing on these tools must ensure that they incorporate the true ethical nature of consent into their designs, allowing for a consenting mechanism that shows case ethics-by-design.

From a historical and evolutionary point of view, it is interesting to note that the creation of the Belmont Report, on which the Menlo Report is based, was the result of medical doctors blatantly disregarding the consent principles laid down in the Nuremberg Code during the Tuskegee Syphilis studies and after widespread outrage and a public uprising. Although the Belmont Report still protects control and decisional power over people's bodies, in the digital age people need control and decisional power over their personal information which information control and privacy are not covered by the Belmont or Menlo Reports. Perhaps it is time for another public uprising to create a new instrument or at least an updated version of the Menlo report that can, as has been done before, explicitly, and further establish and develop the existing consent principles to provide for the ethical application of consent - now also in the digital age and environment.

References

1. Vollmann, J., Winau, R.: Nuremberg doctors' trial: informed consent in human experimentation before the nuremberg code. BMJ **313**, 1445–1447 (1996)
2. Nuremberg Military Tribunal. The nuremberg code. JAMA **276**(20), 1691 (1996)
3. Novitzky, P., Chen, C., Smeaton, A.F., Verbruggen, R., Gordijn, B.: Issues of informed consent from persons with dementia when employing assistive technologies. In: Intelligent Assistive Technologies for Dementia: Clinical, Ethical, Social, and Regulatory Implications. vol. 2, pp. 166 (2019)
4. Human, D., Fluss, S.S.: The World Medical Association's declaration of Helsinki: Historical and contemporary perspectives. In: World Medical Association, pp. 1–24 (2001)
5. Gefenas, E., Tuzaite, E.: Persons without the capacity to consent. Handbook of Global Bioethics. Dordrecht: Springer Science+ Business Media, pp. 85–103 (2014)
6. Beauchamp, T.L., Childress, J.F.: Principles of Biomedical Ethics, 7th edn. Oxford University Press, New York (2013)
7. Brandt, A.M.: Racism and research: the case of the Tuskegee syphilis study. Hastings Cent. Rep. **8**(6), 21–29 (1978)
8. The Belmont Report. Ethical Principles and Guidelines for the Protection of Human Subjects of Research. The National Commission for the Protection of Human Subjects of Biomedical and Behavioral Research (1979)
9. Bailey, M., Dittrich, D., Kenneally, E., Maughan, D.: The menlo report. IEEE Secur. Priv. **10**(2), 71–75 (2012)
10. Lizar, M., Hodder, M.: Usable Consents. Tracking and Managing Use of Personal Data with a Consent Transaction Receipt. UbiComp. Seattle, WA, USA (2014)
11. Open Notice. http://opennotice.smartspecies.com/about/. Accessed 14 Jun 2022
12. Sanchini, V., et al.: Informed consent as an ethical requirement in clinical trials: an old, but still unresolved issue. An observational study to evaluate patient's informed consent comprehension. J. Med. Ethics **40**(4), 269–75 (2014)

13. Kasireddy, P.: How does Ethereum work, anyway. Medium. http://www.easygoing.pflog.eu/32_blockchain_P2P/ethereum_blockchain.pdf. Accessed 03 Aug 2022
14. Consent Receipt Specification 1.1.0. Kantara Initiative Consent & Information Sharing Work Group. Kantara Initiative Technical Specification Recommendation (2018) https://kantarainitiative.org/download/7902/. Accessed 03 Aug 2022
15. Vitor, J.: Towards an accountable web of personal information: the web-of-receipts. IEEE Access **8**, 25383–25394 (2020)
16. European Commission. Regulation on the protection of natural persons with regard to the processing of personal data and on the free movement of such data, and repealing Directive 95/46/EC (Data Protection Directive). General Data Protection Regulation (GDPR), Article 12
17. ISO/IEC 29184:2020 standard for Online Privacy Notices and Consent
18. ISO/IEC 29100 Lead Privacy Framework
19. European Commission. Regulation on the protection of natural persons with regard to the processing of personal data and on the free movement of such data and repealing Directive 95/46/EC (Data Protection Directive). General Data Protection Regulation (GDPR), Articles 9 and 29
20. Santos, C., Bielova, N., Matte, C.: Are cookie banners indeed compliant with the law? Deciphering EU legal requirements on consent and technical means to verify compliance of cookie banners. Technol. Regul. **2020**, 91–135 (2020)
21. Obar, J.A., Oeldorf-Hirsch, A.: The biggest lie on the internet: ignoring the privacy policies and terms of service policies of social networking services. Inf. Commun. Soc. **23**(1), 128–147 (2020)
22. Urban, T., Tatang, D., Degeling, M., Holz, T., Pohlmann, N.: Measuring the Impact of the GDPR on Data Sharing in Ad Networks. In: ASIA CCS. ACM, Taipei, Taiwan. 15 (2020)
23. Consumer Privacy Act. US. Section 1798.120. Right to opt-out of sale of personal information, selling minors' personal information (2020)
24. The National Commission for the Protection of Human Subjects of Biomedical and Behavioral Research. The Belmont Report: Ethical Principles and Guidelines for the Protection of Human Subjects of Research. 1979. https://www.hhs.gov/ohrp/sites/default/files/the-belmont-report-508c_FINAL.pdf. Accessed 27 Jan 2023
25. European Commission. Regulation on the protection of natural persons with regard to the processing of personal data and on the free movement of such data and repealing Directive 95/46/EC (Data Protection Directive). General Data Protection Regulation (GDPR), Recitals 58 and 39
26. Klenk, M., Jongepier, F. (eds.) The Philosophy of Online Manipulation, Routledge (2022)
27. Yu, T.K., Lin, M.L., Liao, Y.K.: Understanding factors influencing information communication technology adoption behavior: the moderators of information literacy and digital skills. Comput. Hum. Behav. **1**(71), 196–208 (2017)
28. Nouwens, M., Liccardi, I., Veale, M., Karger, D., Kagal, L.: Dark patterns after the GDPR: Scraping consent pop-ups and demonstrating their influence. In: Proceedings of the 2020 CHI Conference on Human Factors in Computing Systems, pp. 1–13 (2020)
29. Court of Justice of the European Union. 2019b. Case C-673/17 Planet49 GmbH v Bundesverband der Verbraucherzentralen und Verbraucherverbände – Verbraucherzentrale Bundesverband e.V. ECLI:EU:C:2019:801 (2019)
30. United Kingdom. Information Commissioner's Office. Guidance on the use of cookies and similar technologies (2019) https://ico.org.uk/for-organisations/guide-to-pecr/guidance-on-the-use-of-cookies-and-similar-technologies/. Accessed 27 Jan 2023
31. Brandt, A.M.: Racism and research: the case of the Tuskegee Syphilis Study. Hastings Cent. Rep. **1**, 21–29 (1978)

32. Krugman, S.: The Willowbrook hepatitis studies revisited: ethical aspects. Rev. Infect. Dis. **8**(1), 157–162 (1986)
33. World Medical Association. WMA Declaration of Helsinki – Ethical Principles for Medical Research Involving Human Subjects. Adopted by the 18th WMA General Assembly, Helsinki, Finland, June 1964 and as amended by the 64th WMA General Assembly, Fortaleza, Brazil, October 2013
34. Arachchilage, N.A.G., Love, S.: Security awareness of computer users: a phishing threat avoidance perspective. Comput. Hum. Behav. **38**, 304–312 (2014)
35. Kaspersky.com. One-in-50 online transactions in the banking and e-commerce sectors, were fraudulent in 2019. (2020) https://www.kaspersky.com/about/press-releases/2020_one-in-50-online-transactions-in-the-banking-and-e-commerce-sectors-were-fraudulent-in-2019. Accessed 2023/01/27
36. Baker, S.M., Gentry, J.W., Rittenburg, T.L.: Building understanding of the domain of consumer vulnerability. J. Macromark. **25**(2), 128–139 (2005)
37. Echeverri, P., Salomonson, N.: Consumer vulnerability during mobility service interactions: causes, forms, and coping. J. Mark. Manag. **35**(3–4), 364–389 (2019)
38. McKeage, K., Crosby, E., Rittenburg, T.: Living in a gender-binary world: implications for a revised model of consumer vulnerability. J. Macromark. **38**(1), 73–90 (2018)
39. Baker, S.M., Mason, M.J.: Toward a process theory of consumer vulnerability and resilience. In: Mick, D.G., Pettigrew, S., Pechmann, C., Ozanne, J.L. (eds.) Transformative Consumer Research for Personal and Collective Wellbeing, pp. 543–564. Routledge, New York, NY (2012)
40. Dunnett, S., Hamilton, K., Piacentini, M.: Consumer vulnerability: introduction to the special issue. J. Mark. Manag. **32**(3–4), 207–210 (2016)
41. European Data Protection Board. Guidelines 3/2022 on Dark patterns in social media platform interfaces: How to recognise and avoid them. Version 1.0. Adopted on 14 March 2022
42. Lockton, D.: Cognitive biases, heuristics, and decision-making in design for behaviour change. SSRN (2012)
43. Susser, D., Roessler, B., Nissenbaum, H.: Online manipulation: hidden influences in a digital world. Georgetown Law Technol Rev. **4**(1), 1–45 (2019)
44. Fogg, B.J.: Persuasive technologies. Commun. ACM **42**(5), 26–29 (1999)
45. Botes, M.: Autonomy and the social dilemma of online manipulative behavior. AI Ethics **3**, 315–323 (2022)

Author Index

Printed in the United States
by Baker & Taylor Publisher Services